M000117587

Reverse Engineering

Irreligious Answers to Fundamental Questions

Reverse Engineering

GOD

Irreligious Answers to Fundamental Questions

Michael Rothschild

World Scientific

NEW JERSEY · LONDON · SINGAPORE · BEIJING · SHANGHAI · HONG KONG · TAIPEI · CHENNAI · TOKYO

Published by

World Scientific Publishing Co. Pte. Ltd.

5 Toh Tuck Link, Singapore 596224

USA office: 27 Warren Street, Suite 401-402, Hackensack, NJ 07601

UK office: 57 Shelton Street, Covent Garden, London WC2H 9HE

British Library Cataloguing-in-Publication Data

A catalogue record for this book is available from the British Library.

REVERSE ENGINEERING GOD
Irreligious Answers to Fundamental Questions

ISBN 978-981-123-090-5 (hardcover)
ISBN 978-981-123-091-2 (ebook for institutions)
ISBN 978-981-123-092-9 (ebook for individuals)

For any available supplementary material, please visit
https://www.worldscientific.com/worldscibooks/10.1142/12117#t=suppl

Typeset by Stallion Press
Email: enquiries@stallionpress.com

To my creators, Shoshana and Gideon Rothschild, and to my savior,
Dr. Shabtai Versano

Acknowledgment

I'd like to thank Omer Adelman who helped me improve the formulation of my ideas and avoid some potholes resulting from cultural differences of potential readers.

I'd like to thank Adi Abir and Gonen Yissar for having read early releases of all the articles in my blog, driving me to improve the exposition of my ideas.

Special thanks are also due to Amir Rothschild, Ofer Rothschild, and Danny Jamshy who read many articles in the blog as well as some parts of the book, and provided valuable input.

Last, but not least, I'd like to thank all the readers of my blog whose lively discussions helped make the book better than the blog.

Introduction

At age 17, one of my friends became religious. Many years later, when I asked why he did it, he said he wanted to get answers to questions like "What is my function on this planet?", "What is morality?", "How did life on Earth start?", "How was man created?", "Do I have free will?".

Pondering his answer, I realized that the education system in which he studied, like almost all secular education systems, not only failed to answer these questions, but completely ignored their existence and importance. Only religious education addressed his questions. This understanding motivated me to write a blog, with two objectives in mind: to put together a collection of humanist, atheist, and logical answers to questions that people tend to turn to religion to have answered; and to strengthen people's faith in the power of scientific and rational thinking. My aim was to increase people's ability to distinguish between truth and falsehood, and consequently to discern the benefit of answers based on logic over those based on belief.

Most of the questions that the blog sought to answer have received several answers over the years. I included in the blog only those that seemed to me correct. Some were my own answers. The blog was originally written in Hebrew, against the background of Israeli reality. As the material accumulated, I decided that it would make sense to translate into English the content that is universal in nature and distribute it as a book.

The result is the book you are now reading.

Organization of the Book

In accordance with its two aims, the main content of the book is divided into two parts. The first part may be reminiscent of Kipling's *Just So Stories*, because it contains a collection of answers to separate loosely connected questions. The common denominator that holds the chapters together is the fact that they are sufficiently important for humans to have invented God to answer them, or that they concern fundamental issues that have preoccupied scientists and philosophers from time immemorial. Much of the first part is devoted to thoughts on evolutionary theory, not out of any preference for this field of knowledge over others. It stems from the fact that many of the questions we ask ourselves concern our very humanity, and the answers are based on processes that have led to this humanity. These processes are mostly evolutionary. In this sense, it may be said that in this book, evolution plays to a large extent the roles that religions have ascribed to God.

The second part seeks to reinforce the readers' tendency toward using logical and critical thinking, mainly by presenting solutions to paradoxes. It does so to illustrate the fact that reality does not tolerate contradictions, and that the proper way of dealing with apparent contradictions is to identify the errors that led to them, and not to adopt them and deal with the cognitive dissonance later.

The table of contents classifies the stories into chapters by topic and allows a systematic perusal of them all. Each chapter also contains an explanation of the logic behind the grouping of stories within one chapter.

Links

You will find two types of links in the book:

1. Internal links by which it is possible to go from the table of contents to the various chapters and stories, or from one chapter or story to another, as needed.
2. External links to network resources that are outside the book.

You can distinguish the two types of links by the fact that internal links reference the name of the story and its number. You can read the book without following the links, but you may occasionally use a link to broaden the picture or find a reference. You can find all the external links in the file https://bit.ly/37ZicEC, which has been created for the readers of the paper edition, but online readers may want to use it as well whenever a link in the book is broken. In the Kindle edition, when reading without a network connection, only the internal links will work.

Contents

Part 1
Just So Stories

Chapter 1

Soul, Mind, Will, and Self (Are) Matter

The philosophical difficulty encountered when trying to describe the relationship between mind and body (the mind–body problem) stems from the apparent incompatibility between the subjective feelings and desires of the mind and the objective processes of the physical world. Several theories have been formulated to explain how physical events affect feelings in the mind and how desires of the mind affect the physical world. The various theories can be categorized according to the assumption they make about the relation between mind and body. These include:

1. Dualism: mind and body are two different things. Feelings (qualia), thoughts, and decisions take place in the mind, and the brain is just a mediator between the mind and the physical world.
2. Physicalism: the mind is just a particular organization of matter in the brain.
3. Idealism: only the mind really exists, and matter is an illusion.
4. Neutral monism: both mind and matter are the expression of a third substance.
5. Panpsychism: consciousness is a fundamental feature of the world that exists everywhere, even in elementary particles. The consciousness of complex objects emerges as a combination of the consciousnesses of its constituents. This is an approach I'd like to discard right away.

3

Many people criticize it as "not even wrong," mainly because it doesn't offer an explanation for the way the consciousnesses of the constituents can be "combined," but I'd like to point out that at least the most straightforward interpretation of this idea[1] can be refuted: a person under general anesthesia is a combination of a person and some drug, where the person (who is a constituent) is unconscious.

My approach is that of physicalism. In the following three stories about lost souls, free will, and qualia, I show that this is the only approach that is consistent with the findings of scientific research and with logic.

Some thinkers who embrace physicalism infer that the sensation of self is an illusion.

In the story about the self, I examine their arguments and explain why this sensation represents a real thing.

1.1 Lost Souls

Many religions and traditions speak of the survival of the soul and its reincarnation. Does the soul really survive and migrate?

Mediums of various types claim that they can establish connections with the souls of the dead, and charge money for advice and information about close and distant relatives who had passed on. Do the mediums truly provide the service they purport to deliver? Is there any basis for the assumption underlying these claims, according to which the soul has a separate existence from that of the body?

Many people believe in reincarnation and in other phenomena that rely on a separate existence of the soul. But even those who do not believe in the reality of these phenomena, generally say that their lack of faith stems simply from skepticism and the fact that no one has ever shown them something that would convince them of the truth of these claims. Few people point out that all these claims contradict information already in their possession.

---

[1] Where the consciousnesses of the constituents are not canceled out in the combination.

To address the question of the separate existence of the soul from the body in earnest, we must define what it is that we call soul, or at least specify several of its attributes. Let's consider what these attributes might be.

1.1.1 *Memory*

Most people agree with the argument that the soul should have a memory. Subjectively, memory is the only thing that links who I was yesterday with who I am today. Introspection should lead each person to conclude that it is the memory that accounts for his subjective sense of continuity. It is not surprising, therefore, that most of the apparent testimonies of reincarnation or communication with souls beyond the grave are based on the soul remembering something from the period when it resided in the body of a human known to us. This memory may be a foreign language that the person currently inhabited by the soul knows although he has never heard it, the location of an object that the medium reveals after communicating with the dead, which is not known to any of the people present, or some event that none but the deceased could have known in precise detail.

If we can disprove the fact that the soul remembers, we will deny the basis underlying all these apparent testimonies. We subjectively identify the continuity of existence with memory, so no one is interested in a soul that has no memory. It is clear, after all, that the atoms that make up our body continue to exist in nature and become incorporated in the bodies of animals that eat these atoms after we die, and perhaps even reach, at the top of the food chain, some other humans. But this kind of reincarnation is of no interest to anyone.

1.1.2 *Traits of character*

Some may argue that there is a meaning to the soul, even if it does not preserve memory, because it carries character traits, such as kindness, irritability, honesty, restraint, and other characteristics of human behavior.

1.1.3 *Capabilities and talents*

Some may claim that the wonderful abilities of this or that person in the field of mathematics is the result of the person being the reincarnation of a known mathematician, and so on.

Does it make sense that we have a soul that preserves memory or character traits or abilities after the body dies? We now have a great deal of information that makes it almost certain that the answer to all the above questions is negative. The evidence on which I base this claim is ample, and unfortunately, it arises from functional difficulties we encounter following accidents, surgery, illness, hallucinatory drugs, and more. Consider Alzheimer's disease, which gradually destroys a person's memory, abilities, and character traits. Loss of memory can go so far that he will not recognize the people who were the center of his healthy life: his closest family members. Loss of abilities can leave the person completely helpless. And a courteous and pleasant person can become irritable and aggressive. All these phenomena occur while the person is still alive. Whoever believes in the existence of a soul that does not depend on the body must agree that his soul has not yet departed.

Various accidents can also lead to loss of memory, of various abilities, and a change in character. Memory loss can be partial or full, and certain abilities can disappear whereas others remain intact. I knew a man who, after a stroke, had lost all numeric ability (he was not able to add two and two), without losing the ability to reasonably discuss any other topic. A kind, loving, and thoughtful family man can become a wild, inconsiderate gambler, who destroys all his family connections.[2] It turns out, therefore, that any attribute that we may recognize as a trait of the soul disappears when the brain is damaged, although the soul has not left the body. Doesn't this mean that the attribute, in fact, is not a trait of the soul but of the brain?

We may try to avoid this conclusion in all kinds of ways; I will not discuss all of them here. I address only the most common evasions based on the claim that memory (for example) resides in the soul, but the brain is needed to "retrieve" and express it. This argument ignores the fact that

---

[2] See the sad story of Phineas Gage.

it is possible to "spoil" memories selectively by damaging appropriate parts of the brain. Does this mean that every memory has a unique expression both in the brain and in the soul? Apart from the fact that this would be a kind of "waste" that we never encountered anywhere else in nature, this also suggests that the medium's brain is not suitable for retrieving information from the soul of the departed person, and that the brain in which that soul is later reincarnated is also unsuitable for reading the memories stored in it.

Note that the method for determining the factors responsible for individual's various features attributed to the soul is by and large identical to the method often used to identify the function of genes. Often a particular gene is "broken" to see what defects are caused by the damage. The gene that has been broken is then identified as one of the contributors to the feature that was damaged. The analogy with "malfunctions" in the brain that cause defects in the "soul" is quite clear. Arguing that brain defects are not indicative of the fact that the damaged part (when not damaged) contributes to the creation of the lost attribute is tantamount to rejecting the claim that a defective gene is involved in the creation of the traits attributed to it.

I would like to shed light here on yet another aspect of faith in the preservation of the soul: the moral aspect.[3] Many religions use life in the next world as justification for activities in this one. This is one way in which they encourage believers to act in ways that the religions describe as "moral." Belief in the preservation of the soul is the main reason why some people are willing to waste the only life granted to them (and to others) in this world. Suicidal terrorists are one prominent example of this phenomenon.

1.2 Breaking Free From the Illusion of Free Will

The question whether our will is free has perplexed scientists and philosophers since the days of the Greek stoics.

---

[3]This aspect has nothing to do with the veracity of the claims that religions make about the soul. I included it merely to show that they are not only wrong, but that the implications of believing them can be dangerous.

Our knowledge has now reached a level that enables us to answer this question and deal with the implications of the answer.

1.2.1 *What is free will?*

Some researchers use the involvement of consciousness in the creation of will as a litmus test for the freedom of will. Although this test is interesting, it does not provide us with an answer to the question most of us mean when asking whether our will is free.

The one who sowed the seeds of the interpretation of "free will" as "a will that is determined by consciousness" was Benjamin Libet, who tried to refute the existence of free will through a timing experiment.[4] The experiment showed that a person's choice can be identified by the activity of his brain, even before the person becomes aware of that choice. If this is the case, Libet argued, it is impossible to claim that the will is free: it is not reasonable to define the products of the subconscious as those that "we" have chosen, because by the expression "we" most of us mean "our consciousness." I think most of us agree with the argument that the involvement of the consciousness in our choice is a necessary condition for free will to exist, but some of us ignore the fact that although necessary, it is not sufficient because consciousness itself is not necessarily free. If consciousness is the product of the brain, which is a physical organ that operates according to the laws of nature, the will that this consciousness creates is also subject to the laws of nature, and it is therefore not free.

Unawareness of this insufficiency has prompted many experiments that tried to "save" free will by refuting Libet's conclusions.[5] The purpose of these experiments was to show that, at least in some cases, consciousness is after all involved in the decision process, because it can "veto" what the subconscious proposes to it. The results of these experiments are instructive because they teach us a few things about how the brain works, but they are based on two seemingly erroneous basic assumptions.

The first assumption is that involvement of consciousness is sufficient for free will, which, as we have seen, is not the case: it is necessary but

---

[4] Libet's experiment.

[5] Some of which are described in Neuroscience and Free Will Are Rethinking Their Divorce.

not sufficient. The second assumption is that an experiment is needed to show that consciousness affects the will. The fact that consciousness affects the will must be obvious to all who accept the theory of evolution. For consciousness to be promoted by evolution (and our existence is evidence of the fact that consciousness has been promoted by evolution), it must affect (or be a necessary byproduct of something that affects) our actions in a way that increases our ability to survive, and to influence our actions, consciousness (or its primary cause) must affect our will.

These experiments, therefore, confirm a claim that on the one hand is almost self-evident, and on the other does not attest to the existence of free will.

1.2.2 *Answering the question*

Having understood the question, we can now try to answer it. The answer also solves the mind–body problem.

Let's begin by showing that the brain is indeed a machine that obeys the laws of nature and leaves no room for the intervention of mystical entities like the soul.

I don't review here the vast literature on this subject, or the results of all the research dealing with it. Instead, I focus on the findings of Desmurget's experiment, and explain why these findings say it all. Desmurget's experiment shows that by electrical stimulation of certain parts in the brain, we can evoke the sensation of volition to do something, for example, raising the right arm. What we create through the stimulus is the volition. As opposed to previous experiments that evoked the action itself, Desmurget's experiment evokes the volition to perform it.[6]

Why is this important? Previous research has already dealt with the expression of various sensations in the brain. The neural activity associated with conscious performance of various activities was measured as well. This body of research yielded the identification of neural activity patterns that have been referred to as the neural correlates of consciousness. Note this carefully crafted expression, characteristic of scientific work. The researchers found neural activity patterns that were highly correlated

---

[6] Desmurget's experiment.

with consciousness (hence "correlates"), but these findings were not sufficient to justify the claim that these patterns **are** consciousness itself.

It could be argued that the neural activity patterns that were found to be correlated with some feeling were merely a transient stage between the activation of the senses and the arrival of the sensory information to consciousness. Similarly, it could be argued that the pattern associated with an action is just an intermediate stage between the conscious decision to perform this action, and performing the action in practice. Because of the possibility of such interpretations, the existence of a non-physical consciousness monitoring and controlling the activity of the brain cannot be ruled out.

Some initial confirmation for the claim that consciousness resides in the physical brain and nowhere else came from Libet's experiments. These experiments show that by monitoring the activity of certain parts in the brain of individuals being asked to choose between two alternatives, we can predict their choice even before they are aware of it. Although Libet's results are persuasive, they are not sufficient to completely rule out the existence of a non-physical consciousness, and various objections have been raised. Some of these can be found under the headings "Reactions by dualist philosophers" and "Timing issues" in the Wikipedia description of Libet's experiments.[7] But I believe that all the objections that were raised against Libet's and other similar experiments in the past are inapplicable to the above-mentioned experiment. It cannot be claimed that what we observe in this experiment is an intermediate stage on the way from consciousness to the motor system, or from the senses to consciousness, because volition is, more than anything else, a state of mind. Therefore, the justification for the term "neural correlate" vanishes. Nor is timing an issue in this experiment.

Desmurget's experiment is a clear demonstration of the creation of volition, which is a state of mind, through purely physical means. The philosophical implications of this experiment on the mind–body problem cannot be overstated. Let me reemphasize: the dualistic model implies two flows of information: a sensory one, which flows from the sensory organs to the brain and then to the mind; and an action flow, which starts with a

---

[7] Wikipedia description of Libet's experiments.

decision in the mind, then proceeds to the brain and from there to the muscles. According to this model, volition, which is the basis for the decisions we make, resides in the mind, at the source of the action flow. Information about volition flows from the mind to the brain (as a neural correlate of volition), then to the muscles. Creation of the neural correlate of volition in the brain results in action, not in a sensation of volition. So much for the dualist approach. Desmurget's experiment shows how volition is created in the brain by physical means, which appears to drive the final nail in the coffin of mind–body dualism (as well as that of monistic idealism and that of neutral monism, leaving monistic physicalism as the only survivor).

1.2.3 *Making peace with the answer*

Mounting evidence from scientific research makes it abundantly clear that our will is the product of our brain, which is a physical organ that follows the laws of physics. No exception to this behavior of the brain has ever been documented. Examples of these scientific results can be found in the previous section.[8]

Despite all this evidence, many people seem to cling to the notion of free will. Others, like Penrose, try to "rescue" free will by relying on the probabilistic nature of the laws of quantum mechanics, but this rescue does not work because a probabilistic behavior of the brain does not make it free. It just makes it less predictable. A decision based on the toss of a coin is not free.

Because all the evidence for the claim that our will is not free is already in place, the present section is not aimed at providing new evidence. The questions I'd like to answer in this section are the following:

1. What is it that makes abandoning the belief in free will so difficult, and how can one overcome this difficulty?
2. What are the implications of the understanding that our will is not free for the legal question of personal responsibility?

---

[8]You may also want to watch <u>Free Will With Sam Harris</u>.

To address the first question, let's first try to answer the question "Why do we perceive our will to be free?" I argue that this perception is almost inescapable, even if our will is strictly deterministic. It is the result of the way in which we discover the limitation of the level of freedom of our various capabilities. Consider the way we discover that our movement is not free. We make this discovery when we realize that although we would like to be able to move freely in three dimensions, we cannot do it. It is usually this gap, between what we wish to be able to do in some domain and what we can actually achieve, that makes us realize that we are not free in that domain. Such a gap is almost an oxymoron in the context of our will, because there is normally no distinction between "what we want" and "what we want to want," and this is why we cannot take our perception of free will as evidence of its existence. Note the word "normally" above. There are special cases in which we understand that we would like to want something else than we actually do. A clear example of such a case is that of a pedophile who asks for treatment. A more common example is the difficulty most of us have when trying to stick to a diet. Such cases provide good evidence of the claim that our will is not free, but most of us prefer to ignore this evidence.

This raises the question, "Why do we prefer to ignore this evidence (and, of course, that of scientific research), and maintain the view that our will is free?" Much of the difficulty in abandoning the notion of free will stems from the positive connotation of the word "free." We simply do not want to let go of something good we believe we have.

The solution to the second part of the first question ("How can we overcome this difficulty?") may lie in the realization that the freedom we imagine to enjoy is not as good as we believe it to be. When we analyze the role of will in our lives, we can reach the conclusion that freedom of will is meaningless at best. The role of our conscious will is to identify a state that best satisfies our basic (and uncontrollable) urges, like the urge to eat, to reproduce, to be appreciated, or to feel that we have a free will. This is the solution of an optimization problem, and from the moment the problem has been articulated, its solution has also been articulated, so that as long as we get the correct solution to the optimization problem, the way to reach this solution should not matter. If this solution can be found by a deterministic mechanism, why should that trouble us? What good will we derive from the addition of freedom to that mechanism? Do we need free-dom to reach wrong solutions? I don't think so!

Having reached this conclusion, it should be easier for us to accept the findings of neurological research, which with every new discovery make the existence of free will more and more unlikely.

The second question concerns the legal implications of the absence of free will. Two considerations apply here. The first is that our will has not evolved to enable legislation, and the need to legislate cannot change the nature of will and make it free. The second and more optimistic one is that the lack of freedom of will does not adversely affect the advantage of the existence of a legal system nor the contents of the laws. This is because even if our will is ruled by a deterministic computation, the data that this computation must take into account include, because of the existence of the law and of law enforcement mechanisms, the odds of being caught and the nature of the punishment. This is known as the deterrence element. It may be argued that our chances of enacting good and effective laws increase when the systems these laws are aimed to influence (people) function in a deterministic (and hence, more predictable) way.

1.3 *Qualia*: The Limits of Understanding

Why do we experience red as we do? Why do we feel pain as we do? And, in general, why do we perceive certain feelings the way we do?

These are all questions about the essence of *qualia*, the subjective experience evoked by our senses and emotions.[9] I am among those who believe that our various feelings are nothing but an expression of appropri-ate sets of actions in our brain. I explained why I think so in "1.2 Breaking Free From the Illusion of Free Will." I believe that the analysis presented in that story rules out any other possibility, but it still does not answer the question "Why are things perceived this way and not otherwise?"

---

[9] I extend here somewhat the concept of *qualia* to include also what is at times referred to as propositional attitudes, because in the present context there is no point in distinguishing between the two terms.

Researchers of consciousness who share my opinion try to find an answer to this question as well.

The title of the present story suggests that we cannot quite answer this question. We cannot explain the way we experience our feelings based on something more basic than the feelings themselves. To understand why I make this claim, we must ask ourselves: "What is 'understanding' and what is the process leading to it?"

At birth, we understand very little. In our first years, and especially in the first year of life, we gradually develop the tools that enable us to later understand more and more things. The process begins before birth, and it is gradually integrated with language acquisition. This is not accidental. The two processes have much in common, and there is a reason why the acquisition of tools for understanding begins shortly before the acquisition of language. In "3.1 Think Before You Talk: The Gestation that Preceded the Birth of Language" I note that our ability to create language is most important for our ability to think complex thoughts because it enables us to define complex concepts based on simpler ones that we already know, allowing us to reach ever more general conclusions.

This is similar to the process by which we gradually understand more and more things by basing new understandings on previous ones. The integration of the two processes is quite self-evident: the more we understand things, the more we give them names and expand our language: "And as imagination bodies forth the forms of things unknown, the poet's pen turns them to shapes and gives to airy nothing, a local habitation and a name."

But consider how all this gets started. We must have some initial understandings that are not based on understandings that preceded them. What are these understandings? I argue that our basic understandings are precisely what we call *qualia* (singular, *quale*).[10] These include the inputs to our senses and our inner feelings, including joy, jealousy, sorrow, and even our innate sense of logic. The *qualia* are the atoms of understanding. When we say that we have understood something, it means that we have

---

[10]The story makes two arguments. The first is that there must be a "foundation" of elementary understandings, of the type that we need not (and cannot) explain in simpler terms. The second argument is that these fundamental understandings are *qualia*.

been able to reduce it to previous understandings, and ultimately to reduce it to some *qualia*.[11]

If we ever understand what the *qualia* are, it means that we will have been able to reduce them to... *qualia*. This looks like a contradiction or like a tautology, as I explained in "10.2 I Am Who I Am: Paradoxes of Self-Reference."[12]

This is perhaps the place to note that when I speak of "understanding" I am talking about the subjective experience of understanding. There is also a different kind of "understanding," which is not associated with a subjective experience of this type. We may identify *qualia* with their representation in the brain,[13] and even understand the evolutionary processes that drove the formation of *qualia*, but we cannot understand why such representation creates in us the sense that it does.

It is clear that there is also room for asking whether we all experience our sensory input similarly, and the answer is — no. Yet, most people I spoke with believe that we all experience things in the same way, even if we cannot prove it. When they try to explain their conviction, they often raise arguments of the type: "The fact is that we all agree that red is red and blue is blue."

This, of course, is a logical fallacy because it is based on people's reports and not on their senses. If I experience red the way you experience blue, and *vice versa*, we will still both know that an apple is red because we learn the names of the colors, and I learn that what I experience in a certain way is called red, and so do you, although our experiences are different. But this is only the logical aspect of the error in reasoning. There are, however, also facts that outright contradict the claim itself. There are

---

[11] Note that in English, the word "sense" has several meanings, including: sense as in sensing, understanding, and logic. The expression "to make sense out of X" means reaching an understanding of X. I think that behind the connection between the apparently different meanings of the word "sense" is an understanding that is similar to mine, even if it is not quite conscious.

[12] Partial understanding is possible. The sense of spiciness, for example, is a special case of the sense of heat, but it is clear that there is a starting point for any series of such reductions: the collection of *qualia* that cannot be explained in terms of others. An explanation of a *quale* that is based on the same *quale* is meaningless.

[13] I'm referring to neural correlates of consciousness.

cases in which it is possible to identify the fact that people experience the input of their senses differently. The most obvious case I know is of the viral photograph of the dress[14] the color of which people could not agree about. Another case, which may be considered pathological, is that of synesthesia, the intermixing of senses.[15]

I hope to have persuaded you that the *quale* of "understanding something" is achieved by reducing this something to some *qualia,* and therefore the *qualia* themselves cannot be explained in a way that evokes the *quale* of understanding.

In his paper, "What is it like to be a bat,"[16] philosopher Thomas Nagel argued that materialist theories of mind omit the essential component of consciousness: what it feels like to be a particular, conscious thing (like a bat). Dennett rejected Nagel's claim, contending that any "interesting or theoretically important" features of a bat's consciousness would be amenable to observation. The dispute between Dennett and Nagel about the validity of reductionism may be the result of their unawareness of the conclusion of this story.

We cannot understand the transition from the neural correlates of consciousness to *qualia,* but we can understand why qualia were promoted by evolution. In many ways, *qualia* are the interface between the unconscious and the conscious. One of their main contributions lies in enabling the generalization of various pattern recognition processes. Think, for example, of temporal patterns in which *quale* A precedes *quale* B. This is the type of patterns that enable conditioning, as in the famous Pavlov's dog reflex or those created in the many variations of the Skinner box.[17] Were it not for *qualia,* a separate neural circuitry would have to evolve to recognize each combination of *qualia.* Qualia make it possible to use the same brain circuitry for all combinations of qualia. Similar interfaces between different layers exist also in man-made automated systems.

---

[14] The dress.

[15] Here is an interview with a woman who suffers from synesthesia, or perhaps enjoys it.

[16] What is it like to be a bat?

[17] Skinner box.

The following image can be used to get a feeling of what we discussed in this story.

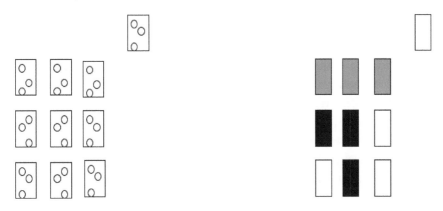

Both parts of the picture contain a pattern that has to be detected in the patterns appearing under it. You can use introspection to discover that:

1. It is much easier to detect the relevant patterns on the right side because we have ready-made qualia for these patterns.
2. To detect the relevant patterns on the left side we must first understand the pattern to be detected, which we do through the general process of reduction to *qualia*.
3. We use the same general process of scanning the patterns in both cases.

1.4 I, Me, and Myself

The "self" is not an illusion

When you ask, "Do I look good?" you refer to your body. This is also the case when you say, "I see myself in the mirror" or "I'm going." Clearly, in this sense, the words *I* and *myself* represent a real thing, not an illusion.

Many scientists and philosophers attribute different meanings to these words. When they talk about the self, they refer to something central, something that feels, wants, and decides, which is located somewhere inside our brains and constitutes our self. Below I will refer to the word "self" in the sense ascribed to it by these scientists and philosophers, and explain why in this sense as well, the self is not an illusion, as many of them claim.[18]

In my opinion, brains that have developed through an evolutionary process **must** contain built-in mechanisms for self-representation. The reason for this is that such brains have evolved because they improve the survival of the organism of which they are part, and to do so, they must know what that organism is. In humans, a region of the brain called the temporoparietal junction appears to be a central component of this mechanism. Recent studies show that this organ plays the same role in other animals as well.[19] Therefore, the self is a real mechanism built into the brain, and not an illusion.

Yet, many scientists and philosophers claim that the existence of the self is an illusion. It is possible to demonstrate that this claim is paradoxical without going into the details of the justifications provided for it, because an illusion must mislead someone, and if there is no such person, there can be no illusion.[20] Nevertheless, I will address some of the justifications various thinkers provide in support of their conclusion that the self is an illusion. These justifications can generate additional insights, beyond those already presented.

I introduce the types of claims made by the deniers of the self by citing the statements of scientists and philosophers. The first statement appears in an article summarizing a book by Sam Harris.[21] The article discusses several issues, but the essence of its position regarding the self is as follows:

The self is an illusion. "What we perceive as self, an unchanging constant experiencer, is really an ever-changing system constructed within the brain." Briefly, what makes the self an illusion according to Sam Harris is

---

[18] When I talk about "something central... that feels and wants" I'm not talking about something that is not physical. My approach is entirely materialistic.

[19] Here is an example of <u>a study showing that the temporoparietal junction plays the same role in other animals as well</u>.

[20] Note that even self-transcendence is paradoxical or meaningless if we understand that it is still necessary to answer the question "Who is the one who transcends the self and feels the beneficial effects usually associated with this activity?"

[21] <u>Article summarizing a book by Sam Harris</u>.

the fact that it is a constantly changing system rather than some fixed point in the brain. But is constant change turning something into an illusion? Is a river, whose molecules, at every one of its points, are continually changing, as waves move across its bed, an illusion? I don't think so, and I suppose Sam Harris doesn't think so, either. Indeed, almost everything in the world is constantly changing, so it is correct to argue that the sense of permanence of something is an illusion, but this is not a claim that characterizes the self.

The article further quotes Harris as saying: "Now that sense of being a subject, a locus of consciousness inside the head is an illusion. It makes no neuro-anatomical sense. There's no place in the brain for your ego to be hiding." Harris appears to ascribe to the proponents of the existence of the self a thought they do not have, and which makes no sense at all (and not only neuro-anatomically). Saying that "the self is in the brain" does not mean that it resides in a specific area of the brain. If this were the case, the same logic would lead one to ask in which part of that area the self resides, and to continue asking the question recursively until pointing to a single atom, and perhaps beyond. It is clear, however, that no one thinks that the self resides in a single atom. Whoever thinks that the self resides in the brain refers to some complex structure in the brain, perhaps even the brain as whole.

The second statement appears in the book *The Meme Machine* by Susan Blackmore. The author argues that the self is a special collection of memes[22] (a *memeplex*) about the correctness of which we are eventually persuaded.[23] She uses the term *selfplex* to designate this special memeplex. But it is inconceivable that in our minds, and in the minds of other animals, there should be a special part devoted to a particular collection of ideas. If the selfplex needed the temporoparietal junction for its existence, it could not affect the genes to lead to its formation, unless

---

[22] The term "meme" has been coined by Richard Dawkins to represent the mental world equivalent of a gene in the biological world. A meme is an idea or thought that can mutate and replicate through its adoption by new individuals. Memes, like genes, are subjected to "natural" selection.

[23] When talking about a meme or an idea, we talk about something that is not genetically imprinted in us. When a trait is inherent in us genetically, this usually means that at some point it has contributed to the physical survival of the creatures who possess it.

the existence of the selfplex granted the organism a survival advantage. This is difficult to believe, given that the selfplexes of different people are different, and it cannot be that they all provide the same advantage. It is quite likely that the selfplex could not have been created in the first place because all of its survival as a collection of memes is based on the fact that, from the outset, the memes that comprise it are very important to the individual in whose brain they reside, and this importance is granted to them by the self. Moreover, I don't believe that even Blackmore would claim that memes are involved when the self is feeling hunger or pain.

The only thing that could have happened (and I think it did happen) is that after the self has been created as a result of normal evolutionary pressures, memes hijacked it, at least in part, for their own purposes. I believe that the selfplex indeed exists.

Although the selfplex adopts a set of beliefs and opinions that get along with one another, and the self is truly ready to fight for them, these beliefs and opinions are not the self itself. Blackmore is right in her conclusion that people tend to identify themselves too much with their ideas and opinions, and that they should acknowledge that it is desirable, under certain conditions, to replace ideas and opinions.[24] She is wrong, however, to say that the self is an illusion. The illusion is the identification of the selfplex with the self.[25]

---

[24] Note that the tendency to defend our positions is also a product of evolution, and it was apparently prompted by the fact that in most cases it provides a survival advantage, which may be related to one's position in the social hierarchy. I have often encountered situations in which people continue to defend positions after they have already been persuaded that they are wrong, only to win the argument. Subsequently though, some of these people adjusted their stated position to the one against which they were arguing during the debate, and tried to obscure the fact that in the past they held different positions. Therefore, it is necessary to invest a conscious effort in identifying the conditions under which it pays to change one's opinion. This distinction is also related to what is said in "2.3 The Trail is Wiser Than the One Who Walks It."

[25] This understanding of things makes it possible to release self-transcendence from the paradoxicality of its definition, and to view it as a real phenomenon with a wrong name (a more appropriate name for it would be selfplex transcendence), in the course of which the self transcends the selfplex.

The effect of the selfplex on society has been massive, as described in "Chapter 5. The Selfplex Rules the World".

1.4.1 *A few associations*

1.4.1.1 *First association*

In the background of Blackmore's remarks is the more general question: how do we have feelings at all (not only the sense of "the self" but any feeling). I addressed this question in "<u>1.3 Qualia: The Limits of Understanding</u>."

1.4.1.2 *Second association*

Some experiments try to distinguish between self-conscious creatures and those that lack self-consciousness by placing them in front of a mirror and checking whether they understand that the stain that appears on their reflection in the mirror is actually a stain they have impressed on themselves. These experimenters do not appear to me particularly smart, and not only because they would classify a blind person or an animal for whom vision is not a key sense as lacking self-consciousness. These experiments ascribe to the animal being tested the ability to identify patterns at a level that they most likely do not possess. Such ability is necessary both to understand what a mirror does in general, and to identify the congruence between the movements of their reflection in the mirror and their own movements. The inability to recognize oneself in a mirror is not necessarily a disadvantage, however. In a popular YouTube spoof, the mirror in a lady's room was replaced by a transparent glass. On the two sides of the glass were identical rooms, and twins were ostensibly making themselves up in front of it, mirroring each other's gestures. An unsuspecting woman (the mark) walked into the lady's room to adjust her make-up, and was shocked to find that she could not see herself reflected in what appeared to be a mirror. Knowing how mirrors function, she was not able to explain her non-reflection. The animals in the above experiments would not fall for this spoof, however, because they lack the ability to understand how mirrors function.

1.4.1.3 *Third association*

The insight that the self is a necessary consequence of evolution has a significant bearing also on how I perceive the threat that might arise from the development of thinking machines, because such machines are not necessarily the result of evolution based on a struggle for survival. Many scientists and futurologists fear that the development of thinking machines can lead to a revolt of the machines against humans. But this result is not necessary if we succeed in creating (without evolution or through a different kind of evolution) machines without a self, or at least without attributes that are based on a self and that lead to aggression. This issue is discussed in the "7.2 Will Frankenstein Rise Against Its Creator?"

Chapter 2

A Brief History of Human Kindness: Empathy, Altruism, and Morality

In this chapter, I propose a way to define morality and explain the evolutionary infrastructure that allows us to adopt this proposition, and which probably excludes other definitions.

2.1 About Morality

"What is morality?" The question is at the center of many debates that are seldom settled, mostly because neither party bothers to think about the meaning of the question.

When the debaters are believers of different faiths, the disagreement is clear because they derive the meaning of the word "morality" from the laws of the different religions. But when atheists are involved in the debate, they don't have such laws to rely on, and the question about the meaning they ascribe to the word "morality" must be answered.

Before addressing this question, it is important to understand that the combination of syllables of any word does not have an intrinsic meaning of its own. The meaning of a word is what speakers of a language **decide** it should be. Therefore, this story does not deal with the question "What **is** morality?" but with the question, "How **should we define** the meaning of 'morality' in a way that serves us best?"

An objection may be raised about the need to define morality in the first place. The opinion that morality is relative and culture-dependent, and has different meanings for different people, is quite frequent, but I have never encountered a person acting accordingly in practice. Quite the contrary: people who proclaim it generally rebuke those who don't agree with them, claiming that their opponents' desire to impose their own definition of morality is, shall we say... immoral?

Contrary to the religious approach, which derives morality from the religious laws, for atheists, the law is in many ways a method of formalizing and imposing their moral understandings. Thus, the decision not to define morality pulls the rug from under the law, amounting to "might makes right."

It is therefore important for atheists to define morality. For the definition to be broadly accepted, it must rely on something that is universal, not relative or culture-dependent. But who is to decide what is moral when "man is the measure of all things?" Man, of course. But again, which man? Isn't it true that every human being has a different sense of morality? Doesn't this path lead back to relative morality?

We all are products of the same evolutionary process, and we all have similar internal notions of morality and conscience. It is no coincidence that when an Orthodox Jew tries to persuade me that religion is the source of morality, he always mentions, in support of his claim, the commandment "Thou shalt not kill," and never the commandment "Kill the homosexual." There is a reason for this pattern. The religious person cites "Thou shalt not kill" because he knows that, although I'm an atheist, I also consider this rule to be moral. How does he know what I think about this rule? I believe that deep inside, although unconscious of this fact, he knows that his judgment of this rule as moral does not stem from his religious feeling but from his innate understanding of morality that all humans share. For the same reason, he doesn't cite "Kill the homosexual" because he doesn't think that I accept it as an example of a moral rule. Quite likely, he too, at some unconscious level, knows that it is immoral.

All humans have a similar conscience, and many features of human conscience can be shown to exist in other animals as well. Human morality is the outcome of this conscience, which is the outcome of evolution, and of additional faculties we possess because of our developed brain.

How is our developed brain relevant to morality? Because of our exceptional capability to predict the results of actions we consider (a capability that improves with time, as we acquire more knowledge and understanding), we are able to "compute" the way these results would make us feel. For example, there is nothing inherently immoral about rolling a rock downhill, but if we think that somebody could be hit and injured by it, we may conclude that the act is immoral. This is a moral conclusion that results from the capability of our brain to predict the possible outcomes of our actions. This capability may be the advantage humans have over beasts in the moral domain.

Anyone considering morality soon reaches the conclusion that a universal morality, based on our inner feelings, which are common to all mankind, must grant equal rights to all individuals sharing it, for the simple reason that to be accepted by all of them it should consider each of them equally valuable. This is the source of the Golden Rule which demands that people treat others in a manner in which they themselves would like to be treated. Such a rule, in some phrasing or another, is part of many moral systems. The various instances of the Golden Rule in entirely different cultures look like an excellent example of convergent evolution[26] in the world of ideas. The environment in which these ideas are selected is our common sense of morality. The existence of such a convergent evolution is yet another evidence for the existence of such a common sense. Further evidence for the existence of a common sense of morality can be found in a study conducted by O. S. Curry, Daniel Austin Mullins, and Harvey Whitehouse, entitled: "Is It Good to Cooperate? Testing the Theory of Morality-As-Cooperation in 60 Societies."[27] The authors propose seven likely candidates for "universal moral rules" — help your family, help your group, return favors, be brave, defer to superiors, divide resources fairly, and respect others' property — which were observed in the majority of cultures studied.

A common sense of morality in humans seems to have evolved from a similar sense of morality observed in other primates and mammals. Frans de Waal has shared on TED some videos of behavioral tests,[28] which show many

---

[26] Convergent evolution.
[27] Is It Good to Cooperate? Testing the Theory of Morality-As-Cooperation in 60 Societies.
[28] Here you can find the videos showing moral behaviors in animals.

of the moral traits all of us (humans and other mammals) share. In one particularly captivating experiment, he shows two capuchin monkeys trained to give the experimenter a stone in exchange for a treat. The monkeys were placed in neighboring cells. When the treat was a cucumber for both monkeys, all was well, but when the experimenter started giving one of the monkeys grapes (the monkeys prefer grapes over cucumbers) instead of a cucumber, the monkey that kept receiving the cucumbers became angry and stopped cooperating. When the same experiment was conducted with chimps, in some cases, the chimp who received the grapes also stopped cooperating, until the experimenter started giving grapes to the other chimp as well.

I suggest defining morality as the conscience we inherited through evolution and the logical inferences we base upon this conscience. The different views of morality in various cultures and religions can be interpreted as different attempts to describe the same **real** phenomenon. The only problem is that once the laws of a religion are cemented, they are almost never revised to reflect the better understanding of the world that humanity has acquired since their foundation.

A fundamental trait that any creature with moral capability must be equipped with is empathy, because it is empathy that enables us to understand what others feel, which is necessary for observing the equality of rights and values mentioned above. In the next story I explain how empathy might have evolved.

Additional evidence for the claim of animal conscience can be found in the following videos:

Hero dog saves another after it was hit on the highway[29]
Adoption among chimps[30]
Sympathetic hippo[31]
Battle at Kruger[32]
Leopard and a baby baboon[33]

---

[29] Hero dog saves another after it was hit on the highway.
[30] Adoption among chimps.
[31] Sympathetic hippo.
[32] Battle at Kruger.
[33] Leopard and a baby baboon.

A herd of elephants rescues a baby elephant[34]

You are also encouraged to watch the following related videos:

The Great Debate Panel[35]

Is God Necessary for Morality? William Lane Craig vs. Shelly Kagan Debate[36]

And read Michael Shermer's book *The Science of Good and Evil*.

2.2 Why Do We Empathize?

The topic of altruism, a behavior that serves others and not the individual displaying it, has always attracted the attention of evolutionists and of deniers of evolution alike.

Evolutionists have explained altruistic behavior by game theory considerations and as byproducts of kin selection. I believe these explanations are correct, and I don't repeat them here because everyone interested in the subject knows them already. This story exposes an additional perspective on the subject, which relates directly to our sensations, not only behaviors (behavior can be explained without relying on intentions).

Humans are equipped with a large brain, similarly to mammals in general and birds. One of the main tasks of this brain is to contain a model of the world, which we use to predict the consequences of our intended actions and spare ourselves the execution of actions that might harm us. The physical information serving as input to the model comes through the senses, and can be easily combined with models of the physical world to predict, say, the expected trajectory of a stone. But simple models of this type are not sufficient to predict the behavior of complex living beings, for which we must understand what they experience, that is, we must be able to identify in some way their inner feelings.

The problem with feelings (*qualia*) is that we experience them in a direct way. Our inability to effectively convey these feelings through regular communication has often been characterized by phrases like, "you cannot explain the feeling of the color red to a blind man." Seeing red is an

---

[34] A herd of elephants rescues a baby elephant.
[35] The Great Debate Panel.
[36] Is God Necessary for Morality? William Lane Craig vs. Shelly Kagan Debate.

immediate experience that seems impossible to describe to anybody who has not had it. Even more difficult would be to understand what's going on in the head of someone who doesn't attempt to convey his feeling to us.[37]

How can we understand the other at all under these circumstances? We can do it because we hold a model of others in our brain, which is designed to serve us to predict their behavior, and provide us with approximate information about their feelings. For this model to work, it must make us feel what the other feels. This is the mechanism that causes people to cry when watching a movie, or to feel angry when witnessing an injustice suffered by another. This is empathy.

When we face an unpleasant experience, our natural reaction is to attempt to eliminate its cause. It is therefore equally natural, when the suffering of others creates suffering in us through the model of the other we hold in our mind, that we should seek to ease their suffering to reduce ours. This is altruism.

To summarize: Our minds have a model that simulates the other. This model is needed to predict the behavior of others, therefore its formation has been "encouraged" by natural selection. Because the only way such a model can tell us what the other feels is by causing us to feel nearly the same way (empathy), it makes us suffer when others suffer. An altruistic action is one intended to ease this suffering.

Even more briefly: Empathy and altruistic feelings are necessary byproducts of our ability to predict the actions of the other.

Up to this point, the argument relies strictly on introspection and logic. It looks like "armchair science." But there seems to be abundant evidence for it. The first and easiest to access, almost without having to leave the armchair, is the way movies affect us. Movies turn out to be an excellent laboratory for testing our hypothesis. In real life, because of its complexity, it is difficult to distinguish between the various sources of our sensations and impulses, and our actions stem from combinations of many feelings

---

[37] This point should be clearer after reading "1.3 *Qualia*: The Limits of Understanding." Our own understanding is based on our qualia, and to understand what others feel, we must somehow get hold of their qualia, which we can only feel, and not understand.

and impulses. In movies, we encounter characters toward whom we have no prejudice. We owe them nothing, we don't depend on them, and are not intimidated by them. All our feelings toward these characters are the result of our identification with them, or of what they may have caused the characters with whom we identify. In movies, our experiences of joy, sorrow, pain, love, anger, and fear of others are refined and distilled.

The second evidence is provided by mirror neurons. About 20% of the neurons involved in any of our actions or sensations seem to also be activated when we watch others performing these actions or having these sensations.[38] Mirror neurons have been found in humans and in monkeys, but I tend to believe that other animals also have them. Yawn contagion in other animals,[39] such as wolves and dogs, strongly supports this assumption (even cross-species yawn contagion has been observed).[40] According to my interpretation, these neurons are a useful mechanism, the development of which was encouraged by evolution to enable us to get a direct understanding of what other creatures feel — an understanding we would not be able to get in any other way.

Failures in the development and activation of these neurons have often been linked to certain types of autism. People displaying various types of autism are unable to "get into the head" of other people.

In their perception action model,[41] De Waal and Preston described insights that are similar to mine, but they restricted the reasons of the development of empathy to the advantages it provides in interactions within one's own social group. They seem to have stopped one inference too early, because the ability to understand the other is useful also when the other is not a member of one's social group. Moreover, the mere forming of a social group of similar creatures with complex brains and behaviors (as opposed to societies with predefined roles, like those of ants and bees that act as a single organism) may require empathy to begin with.

---

[38] Follow this link to watch Prof. Ramachandran explain experiments he performed to detect the effect of mirror neurons on our sensation and behavior. He also makes some inferences about the origins of these neurons, but he seems to be unaware of the fact that mirror neurons have also been found in monkeys.

[39] Yawning is contagious among animals: study.

[40] Chimps respond to human yawning.

[41] Perception action model.

The following are some notes on the explanatory power of the above description of the evolution of empathy:

1. It deals with the evolution of genuine altruism, which begins with our feelings, and those of other developed animals. Contrary to the altruism that Price's equation explains to some extent as "egoism in disguise," this story describes the evolution of altruism that moves us to help others without contributing to our own fitness, or even to that of our genes. Price himself seems to have understood that his equation does not describe true altruism, which may have precipitated the end of his life. In his book *The Price of Altruism*, Oren Harman argues that Price became a "suicidal altruist" in an act of rebellion against the conclusions of his equation. In other words, had Price been aware of the argument presented in this story, he may not have committed suicide.

2. As opposed to the belief that the evolution of altruism depends on some kind of fortuitous development, in the above description, altruism is a feature that necessarily evolves in creatures that have a brain able to model the behavior of other creatures.

3. The description releases altruism from dependence on the dubious concept of group selection.[42]

4. It shows that altruism does not depend on reciprocity. Note that reciprocity is a case of almost irreducible complexity because it requires a relatively high number of altruistic creatures to sustain itself. This almost makes it a non-starter because the mutation that creates reciprocity-dependent altruism for the first time occurs in only one individual, which will leave on average fewer offspring than its competitors (this individual pays the fee of altruism without harvesting any return on its investment).

5. It explains inter-species empathy and altruism.

6. It explains why we empathize even when we are not required to respond, as, for example, when we feel the joy of the object of our empathy.

---

[42] http://en.wikipedia.org/wiki/Group_selection#Criticism.

2.2.1 *And those who do not empathize...*

If empathy is indeed a byproduct of the capability to "read the mind" of others, how can we explain the existence of psychopaths?[43]

Not all psychopaths are created equal. I think that there are at least two reasons for the appearance of psychopathy, which are consistent with the main idea of the present story.

1. Empathy is the ability to translate the feelings of others into one's own feelings. If one's ability to feel pain or fear is impaired, one cannot feel these *qualia* in others. The existence of the link between fearlessness and psychopathy has been demonstrated in research.[44]

2. Once the ability to translate the feelings of others into one's own feelings is present, individuals may benefit from an enhanced capability to discern between their own pain and that of others. An example of such a benefit is a better ability to manage other people (using other people as tools). Studies have shown much higher proportions of psychopathic traits in high-level managers than in the general population.[45]

Does the second reason mean that we should expect to eventually become a Klingon-like[46] psychopathic society? I'm cautiously optimistic about the answer to this question because I believe that our organization as a society has driven us beyond the point of no return. A legal system, with enforcement and punishment mechanisms, turns any potential advantage of excessive psychopathy into a disadvantage.

Here are some relevant Internet resources:

---

[43] Psychopathy is a personality disorder characterized by **impaired empathy**, disinhibited and egotistical traits, and persistent antisocial behavior.

[44] How Fear Makes You Do Good Or Evil.

[45] Proportions of psychopathic traits in high level managers are 3 to 21 times larger than in the general population.

[46] The **Klingons** are a fierce and aggressive human-like alien species in the *Star Trek* science fiction series.

Feeling your pain by Bonnie Prescott[47]

Empathy hurts by Marco L. Loggia, Jeffrey S. Mogil, M. Catherine Bushnell[48]

I know how you feel by Stephanie D. Preston and R. Brent Stansfield[49]

Social Bonds and the Nature of Empathy by Watt, Douglas F.[50]

Some videos showing empathy and altruism in various animals:

The elephant and the dog[51]

The orangutan and the hound[52]

A dog risks its life for another[53]

2.3 The Trail is Wiser Than the One Who Walks It

This adage is familiar advice to hikers, intended to encourage them to walk along the paths that have already been trampled by many feet, and not risk venturing on new paths. I wish to use it as a metaphor, and liken humanity to a walker, and the many generations in its evolution to the people whose feet have paved the path.

I do not intend to argue against innovation, and often the wise action is to deviate from the common path, but such action is truly wise only if the reasons that have led to the paving of the common path have been understood.

In recent generations, disrespect for our natural impulses, especially those having to do with negative emotions such as anger, hatred, jealousy, insult, etc., has become an accepted approach that often justifies violence (at least verbal) against those who exhibit such impulses. This approach is not without foundation, if it is adopted prudently and if it takes into account the reason for the development of these impulses, the reason for the advantage of restraining them nowadays, and the circumstances at

---

[47] Feeling your pain.
[48] Empathy hurts.
[49] I know how you feel.
[50] Social Bonds and the Nature of Empathy.
[51] The elephant and the dog.
[52] The orangutan and the hound.
[53] A dog risks its life for another.

hand. But it has been gradually becoming a kind of political correctness that people adopt without discretion.

Evolution would not inculcate in us negative emotions and impulses driven by these emotions unless these feelings and impulses provided some advantage to their owners. At a time when evolution introduced these feelings, renouncing them was dangerous for the survival of the lineage. Those who didn't make sure others understood that they were not pushovers, by taking violent action against those who harmed them, simply invited another hit.

Today it is possible (and often preferable) to suppress the activity that these feelings urge because we have been able to organize as a society that generally provides individuals with better protection than is the one provided by obeying impulses. Note, however, that the advantages that members derive from human society have not eliminated the negative emotions we experience in some situations. Admittedly, evolution has not yet had enough time to do so, but my assessment is that these feelings will remain with us forever, because even in the most reformed society we must still assert our rights.

Quite naturally, some people refer to the behavior that has been inculcated in us by evolution as "emotional." I think this is correct. What I believe is mistaken is the idea that behavior not inculcated in us by evolution is preferable only because it is not emotional.

I found an intriguing expression of this state of affairs in an article in *Scientific American*[54] that I came across. The article raises three issues, and I disagree with the conclusions on all three, but what prompted the writing of the present story is a psychological experiment described in it.

The experiment assumed that people who think about a deterministic world in which desire is the product of a series of events resulting from the laws of nature, have a different attitude toward responsibility for crime and the ensuing punishment in a world of this type, depending on whether they consider the issue emotionally or theoretically/rationally. To test this argument, they assembled a group of people, described to them a hypothetical world in which the will is indeed the result of events dictated by the laws of nature, after which they divided participants into two groups.

---

[54] Experimental Philosophy: Thoughts Become the New Lab Rats.

The members of one group were asked a question that was intended to stimulate theoretical/rational thinking, and the members of the second group were asked a question that was intended to stimulate them emotionally. They did this in several ways, the first of which is the easiest to describe, but all the rest are variations on the same idea. I describe only the first experiment, because the results obtained in the other ones were similar.

In this experiment, members of the first group were asked a simple question: "In a world like this, are people morally responsible for their actions?" Members of the second group were asked a more elaborate question: "In a world like this, a man named Bill is attracted to his secretary and decides that to be with her, he has no choice but to murder his wife and three children. He knows that there is no way out of his house in the event of a fire. Before going away on a business trip, he activates a device that sets the house on fire, causing the death of his wife and children. Is Bill responsible for the death of his wife and children?"

As expected, the answer of members of the first group tended to be negative, whereas that of members of the second group tended to be positive. These results were confirmed by the more complex versions of the experiment.

On the face of it, we have a scientific experiment that proves the basic assumption of the experimenters. My disagreement, however, stems from the combination of one factual claim and two arguments that can each be neatly represented by a well-known maxim.

The factual argument is that the hypothetical world described, which in fact is not hypothetical, is exactly the world in which most scientists think that we live.[55]

The first maxim is "The trail is wiser than the one who walks it." In the metaphor I used, it means that if evolution has produced a certain outcome, we must understand the reason for this result before formulating a theory leading to a different theoretical/rational result that predicts behavior.

The second maxim is "There is nothing more practical than good theory" (attributed to Kurt Lewin). If the theoretical/rational solution

---

[55] 1.2 Breaking Free From the Illusion of Free Will.

reached by both experimenters and participants in the first group is different from the solution found by evolution (applied in practice by legal and enforcement systems worldwide, under which both participants and experimenters live), **there is good reason to believe that the theory** on which they relied in their theoretical/rational thinking is not good.

The combination of these two reasons, in and of itself, should have raised a red flag before the researchers. I explain the results differently. More than it promotes the formation of a true perception of reality, evolution promotes in its creations a beneficial response to the perception of reality that has been formed in them. Thus, even in a situation in which the worldview intuitively created in humans includes a false belief in the existence of free will, evolution promotes behavior that encourages survival.

The behavior promoted by evolution, the one that the participants in the second group could not escape when the concrete case was presented to them in explicit detail, is one of imposing responsibility and appropriate punishment. This suggests the possibility that the discernment of respondents in the first group was incorrect, and there is reason for imposing responsibility even in the absence of free will.

I explain the reason for this at the end of "1.2 Breaking Free From the Illusion of Free Will": "even if our will is ruled by a deterministic computation, the data that this computation must take into account include, because of the existence of the law and of law enforcement mechanisms, the odds of being caught and the nature of the punishment. This is known as the deterrence element. It may be argued that our chances of enacting good and effective laws increase when the systems these laws are aimed to influence (people) function in a deterministic (and hence, more predictable) way."

In other words, the humans' emotional consideration is the result of the better theory that evolution has found. If experimenters and the respondents had expanded their theory to include this insight, their theoretical/rational consideration would also have told them that the murderer must be held responsible for his actions. Their error stemmed from the fact that at some point, in the course of formulating the theory, instead of incorporating the part that explains why personal responsibility exists in a deterministic world as well, they incorporated their feelings that told them that in a deterministic world there can be no personal responsibility. These feelings derive from the belief in the connection between free will and responsibility, a belief that could develop only in creatures who believe in the existence of free will.[56]

These feelings are an example of a place where the walker can indeed be smarter than the trail. After all, the belief in free will has survived natural selection. It is, however, not a result promoted by evolution, but a byproduct of the circumstances, which normally, unlike the present discussion, does no harm.[57] This belief is erroneous, as scientific research shows, and the reason for it is also explained in the first part of "1.2 Breaking Free From the Illusion of Free Will."

Being responsible for one's actions means having to bear the consequences of these actions and if the consequences are a punishment, being responsible means being punishable.

The *qualia* of responsibility and punishability are not the result of a philosophical debate on morality; they are the results of evolution. This becomes abundantly evident when we encounter them in animals.[58] Evolution selected for them because they are beneficial and not because they are moral. As explained in "2.1 About Morality," morality is the outcome of our built-in conscience and not the other way around.

2.4 The Next Steps in Morality

In an earlier story "2.1 About Morality," I talked about the meaning that should be infused in the concept of morality when "God is dead." In this story, I talk about what we should do to expand the meaning of the term beyond what evolution has imposed on us.

---

[56] Intelligent creatures that are aware of the fact that will is not free would realize that responsibility is justifiable without it.

[57] Note that natural selection does not select for true beliefs. It selects only for beneficial behavior. The belief that will is free, when combined with the belief that when will is free, people are responsible for their actions, leads the individual to act in a beneficial way and hold people responsible for their actions.

[58] Some examples of retaliation by animals.

In the story "2.1 About Morality," I mentioned the evolutionary constraints that created our conscience, and talked about expanding the dictates of conscience to morality as a whole, through our ability to predict and reason. Because the computational work on the logical implications of the dictates of conscience is still largely ahead of us, the question arises whether the data at our disposal are sufficient for this computation. In other words, is an in-depth deployment of our predictive capacity sufficient to determine the degree to which the outcome of our activity will satisfy the requirements of our conscience?

We seem to have a problem here because our conscience has evolved on the basis of various impulses, including empathy, and empathy is limited by two constraints:

1. Being based on our sense of ourselves, it can help us understand what is good for others only in matters in which we have a reasonably good understanding of what is good for ourselves.
2. It developed mostly before we became a knowledge-based technological society capable of substantively changing its living environment.

Indeed, there is an alarming interaction between these two constraints. The fact that our technological progress is changing our environment at an increasing rate places us in situations that our emotional evolution has not prepared us for. To continue to expand morality so that it can deal with questions expected to arise in the future, we cannot rely only on what evolution has imprinted on us; we must define a goal for that expanded morality.

We could try to compute what evolution would have imprinted on us if it had continued for many more years, but I believe that this is not the right way of addressing the problem because the demands of evolution are not sacred, and because evolutionary pressures depend on the direction in which we choose to lead society. We have no choice but to accept the dictates of evolution that have already been imprinted on us, but those that have not yet been determined have no special value for us.

We can examine, nevertheless, whether the answers given to questions at this time remain stable over the evolution of the individual and of the culture, so that once they become part of culture, no contradiction is expected to arise between what we formally define as morality and what we feel as human beings. These goals require decision and agreement. For the reasons given above, introspection is not sufficient to determine them. The following is an example of two fundamental decisions that we must make before we can complete the task. We are already confronting the results of not having made these decisions yet:

1. What is the overall importance of life span vs. that of quality of life?
2. Is it important for us that mankind survive the maximum time that planet Earth will allow it, or is it more important for us to derive from planet Earth the maximum human happiness, even if it brings the end of humanity closer?

The first question has to do, among others, with our attitude toward euthanasia, abortion, and possibly drug use. The second concerns the legacy we leave to future generations. This legacy has several aspects: one concerns the resources of planet Earth, another our cultural heritage, and affecting both is our control over the rate of population growth.

There are other similar questions, but I would like to focus attention on a set of problems that we just started dealing with, although we should have dealt with some of their aspects long ago. I'm referring to implications of the rise of artificial intelligence (AI) for morality, which embrace at least the following two domains: the way in which we should relate to work and moral decisions that machines will have to make.

2.4.1 *The way we should address work*

Fewer and fewer people are required to do the work that humanity needs, and at the same time, an increasing portion of the tasks that must still be performed by people require high education and intelligence. To the extent that this development continues, it will result in an increasing proportion of the population being unable to find work because it will be unable to acquire the skills necessary to carry out the work left for human beings.[59]

---

[59] Note that the current situation is different from that in the beginning of the industrial revolution. The industrial revolution replaced only the physical power and speed of humans by those of machines; the workers were still required to use their brains. With AI, we appear to be approaching a time when machines will outperform humans in just about any task.

These circumstances are related to the topic at hand because accelerating technological change leads to a situation in which a small portion of the population will be working and the rest will be forcibly unemployed. At the same time, all the work necessary for the survival of all mankind will be accomplished. Currently, a person must work to earn the money needed to sustain a proper lifestyle. The trend described above, however, will sever the link between the ability to work and the ability to survive.

Breaking this link is not trivial. In Israel, this link has been lost for an entirely different reason: the ultra-Orthodox have sufficient political power to introduce laws that allow them to live without working, at the expense of other parts of the population that as a result must pay more taxes. Needless to say that many working people resent them for this. People are reluctant to pay taxes to support those who don't want to work. It contradicts their sense of fairness. Luckily, they are not as reluctant to pay more taxes to support those who can't work. Those who work will have to be reassured that those who don't are not simply lazy.

In the economic sphere, we will have to build an arrangement based on something between capitalism, which encourages creativity but attaches harsh living conditions to unemployment, and communism (the theoretical one, not the one that has been implemented), which creates excessive equality and therefore encourages idleness and kills productivity. Universal basic income and the Venus project represent two different approaches to the solution of this problem.

I cautiously note that there are other possible methods of compensation, which are not acceptable in the current because they tie together productivity with what we perceive as essential human rights. For example, one's political influence may be associated with one's productivity. I'm afraid that expanding on this topic may lead to my being stoned.

2.4.2 Moral decisions machines will have to make

People make moral decisions all the time, often without even being aware of it. Making moral decisions is part of what it means to be autonomous. Now, as machines are becoming ever more autonomous, they too will have to make moral decisions.

Consider an autonomous vehicle that has lost its breaks, and the AI system governing it has to select between a course of action that will best protect the people inside the car and one that will best protect people on the sidewalk. What should it do? This is a variation of the well-known trolley problem, a thought experiment designed to compare different moral stances in different people. Various versions of this problem have been used in many surveys, and the answers were far from unanimous. The question that comes to mind is: If people cannot tell what the moral behavior should be in such cases, how are they expected to program the machines to react in a moral way?

Is this a show stopper for AI? No more than it is for humans in such situations. The way we should deal with it is the one described in the beginning of "2.4 The Next Steps in Morality."

Note how the development of intelligent machines forces us to take a deeper look at what we consider as "being human."

1. The importance we attach to fairness is what turns the job issue into a problem. In the previous stories we saw that this is the product of evolution, and that humans share it with other species.
2. The central role that working plays in the establishment of the feeling of fairness, and hence the threat to this fairness that arises from the capability of machines to replace us at work.
3. The ties between human intelligence and human morality, which make us want to use our intelligence to extend morality, and use the intelligence of machines to make them understand it.
4. The difficulty we have in defining morality in situations evolution has not prepared us for.

Other aspects of what we see as "being human" are described in Chapter 3.

Chapter 3

Being Human: Capabilities

We are concerned a great deal with questions about the human condition. These questions seek to position us on a scale (imaginative and chauvinistic, but convenient for metaphorical use) that leads from plants, to non-human animals, humans, and finally "God." We'd like to know, on the one hand, what are the special virtues that differentiate us from other living beings and what sets us apart, as living beings, from plants. On the other hand, we'd like to know what differentiates us from God, in other words, what are the limitations imposed on us by the fact that we are living beings created in an evolutionary process?

3.1 Think Before You Talk: The Gestation that Preceded the Birth of Language

In the Torah, even the serpent speaks. But rational people must ask themselves, how it is that humans speak and that they do it in so many languages. (The "Tower of Babel" answer is not particularly persuasive.) I argue that the development of language occurred in the brain and was mostly gradual.

With their characteristic chauvinism, humans keep searching for the answer to the question: "In what way is man superior to the beasts?" The search is based on the assumption that there must be an answer to the question, and that it cannot be that human beings have no substantial

41

advantage over animals. One common answer to this question is "language," that is, the argument that humans are superior to beasts because they have language. I argue that despite the obvious bias in favor of humans that is inherent in the question, a less biased answer happens to be at hand. In other words, man indeed has preeminence[60] over beasts, and "language" is almost the correct reason for it.

3.1.1 *Animal languages*

The leader of a pack of wolves is confronting a contender. He bares his teeth and his hair bristles. The contender chooses to submit, which he demonstrates by a shameful retreat, and life returns to normal.

What happened here is that the two wolves conducted a negotiation, which ultimately saved effort and risk to both of them. The conversation took the form of a series of signals — a collection of gestures and sounds intended to influence the behavior of the interlocutor. Such conversations often take place between animals, even if they are not of the same species, and quite similar conversations exist even between plants.

Are these capabilities, which exist in most animals, the basis of human language? To answer this question, we need to understand how the ability to exchange information came about, and why the transition to a human-like language is not "more of the same thing."

3.1.2 *Biological evolution of communication methods*

A good way to begin understanding the process of biological evolution of communication methods is to pay attention to the fact that some communication exists even in plants.[61] Trees, for example, can respond to certain approaching dangers, like water shortage or the spread of disease, before coming in direct contact with these threats, relying on "signals" they receive from other plants around them. How does this happen? After all,

---

[60] I realize that this word is also biased because it is based on human values, but that's not the point here.

[61] Read more about <u>conversations between trees</u> and about <u>plants "alerting" one another (and themselves)</u>.

the trees did not confer and agree whether and how to signal, or whether and how to interpret signals. What seems to have happened is this: plants facing a danger had reacted in one way or another. Not intentionally, but rather in a chemical process, the reaction resulted in the release of certain substances into the environment, or by the cessation of such release. Plants that "learned" to recognize these changes in the environment and to respond to them appropriately increased their chances of survival, and in this way, without deliberate intent or awareness, these changes became a type of signaling. Signals of this kind exist also in the animal world, many of them based on the sense of smell.

3.1.3 *Communication methods based on pattern recognition*

Consider the following scenario: a predator is preparing to attack. The attack is based on biting, and therefore the predator bares its teeth. Animals that identify the pattern in which after the baring of teeth comes the attack, will learn to recognize the baring of teeth itself as a threat, increasing their chances of evading the attack. Now, if the predator recognizes the pattern whereby following the baring of its teeth, other animals tend to flee, it may start baring its teeth even when it wants only to chase them away. The baring of teeth has become a signal.

The signal development processes described above are simple, and are not the result of any initiative to create a signaling language. Nor do they encourage such initiative, because in all of them, the understanding of the pattern preceded its transformation into a signal. If so, what is behind the creation of human language, in which most of the signals (words) where created deliberately?

3.1.4 *Human languages*

We often come across articles that try to trace the origins of human language. Almost without exception, these articles raise one of two hypotheses. The first is that language originates in the collection of vocal calls and signals that are common among animals. The second is that the source of language is in physical gestures, and that the first human language was sign language, which then evolved into vocal language, in parallel with the development of the voice box in our throat. In his wide-ranging book, *Unweaving the Rainbow*, Richard Dawkins attempts to trace, among others, that formative event in which a genius first invented language. My argument is that all the above approaches miss the main point, which is that the development of language occurred in the brain and nowhere else, and that it was mostly gradual, without the involvement of any genius or formative event.

This argument is based on the following considerations:

If language came into existence as a result of a brilliant stroke of genius on a certain occasion, without preceding development of the brain, all the animals around us would have been speaking by now. After all, they don't need that genius and special event because we already provide them with everything that the genius and the event could have supplied. More than that, despite active efforts to teach them language, our success is very limited.[62] Indeed, this consideration is a special case of a deeper truth, having to do with the development of language as part of the human evolutionary process.

Evolution, as we know, is a gradual process based on mutations that survive and spread in the population as a result of the benefits they grant individuals who carry them. But what advantage could any single individual have derived from the accumulated mutations responsible for our ability to learn languages? Such an individual would not have been able to talk to anyone around him, and the opportunity to converse would have been of no use to him. How, then, did the ability of speech spread to become public property?

The challenge these mutations had to contend with was similar to that facing the best salesman in history, who tried to sell the first telephone when no others were around, so there was no one to talk to with it. The salesman was at least able to offer the device to several people at the same time, suggesting that they talk to each other, an option that was not

---

[62] "Very limited" is different from "non-existent;" some success has been achieved and in rare instances, chimpanzees have even been able to invent new expressions. In one case, a chimpanzee invented the term "drink fruit" to describe a watermelon.

available to evolution, which operates on individuals rather than on whole groups in a coordinated manner.

3.1.5 *Evolving, fast and slow*

Human language developed in two stages: a genetic and a memetic stage.[63] In the genetic stage, the human brain evolved and the ability to conceptualize was created. During this biological evolution, the relevant abilities gradually accumulated and spread through the population because of the advantages they granted their owners in the era preceding spoken language. This stage took a long time, because the passage of many generations is needed for mutations to become widespread in the population, the more mutations are involved, the further the timeframe stretches.

The memetic stage was much shorter because at this point only ideas, not genes, had to propagate throughout the population. There was no need for the passage of generations at all, and the idea of language could reach the entire population in a single generation.

It may be that once the idea of language was established in the general population, an additional stage of biological evolution began, which through natural selection "preferred" people who could speak better. As a biological stage, this could have lasted a long time and may continue into the present.

3.1.6 *Inventing language*

I noted above that language is *almost* the correct answer to the question of the reason for human preeminence over animals. Humans do not only speak one language or another. They can relatively easily acquire additional languages, including spoken, sign, and written ones. The ability to acquire languages is stronger than the ability to speak any given one of them, but even this does not exhaust the essence of our competence in this area, which in my opinion is in our capacity to invent language. Yes, the ability to invent language, not the ability to use it or acquire it. This

---

[63] Meme is a term coined by Dawkins (as part of his ability to invent language) to describe the evolutionary unit of ideas. <u>Read more about memes</u>.

capacity is manifest most poignantly in mathematicians, who often invent words even as they formulate their proofs. Their use of this process is so common that, so as not to strain too hard in choosing the right words, they ascribe entirely new and different meaning to the words they have used in the past, acknowledging that this meaning will serve them only in this particular instance.

Let's take the word *epsilon*. Anyone who has ever dabbled in mathematics knows the opening sentence: "Let epsilon be a small arbitrary positive number so that such and such..." This is in practice a declaration that from this moment onward, until the end of the present portion of the proof, the word *epsilon* will be used to denote one specific thing, without committing to any meaning that might be ascribed to it in other contexts. In the occasional lectures I give on solving puzzles, I bring more interesting examples, but this is not the place to elaborate.[64]

Although the ability to use language does not give the individual any evolutionary advantage, the capacity to invent language does, because the mechanism of linking a complex idea to any symbol simplifies the mental approach to that concept and enables the construction of more complex notions that make use of it.[65] This mechanism, when applied recursively, allows the individual to hone his understanding of the environment, and enhances his ability to survive.

Over time, this advantage to the individual translates into a higher number of offspring, and eventually becomes public property. I think that people began to talk to each other only after most of them had already been "talking to themselves," not in any formal language, but in an internal one that each of them developed for himself.[66]

---

[64] You can find an instructive example in "11.1 Word Power."

[65] After reading an initial version of this piece, my father referred me to a treatise written in 1772 in which similar ideas were presented, except for the evolutionary argument, which could not have appeared before the discovery of evolution. For some reason, most of those dealing with the origin of language ignored this article, and therefore I also have heard about it only recently. See Johann Gottfried Herder (1744–1803), *Treatise on the Origin of Language*.

[66] Some people go as far as claiming that thinking is entirely dependent on language. I think they are exaggerating. Some kind of thinking had to be performed before language existed, otherwise language could not have been invented.

Armed with this idea, we can now delve into the problem Dawkins was trying to tackle: How did it occur to some genius, all of a sudden, that it may be possible to converse with others as well?

As I argued, humans had already acquired some type of internal language that allowed them to define complex terms. What enabled humans to bridge the gap between internal and spoken language was that in addition to recognizing patterns, like all animals, they also conceptualized the idea of pattern recognition and were able to discern patterns according to which other animals recognize patterns. A person who identified such a pattern could begin using it to "train" those around him. He could, for example, tell someone around him, "Get out!" and immediately take aggressive steps to make the other recognize the pattern in which following the utterance "Get out!" comes the attack. He could say "apple" every time someone around him looked at an apple, and make that person associate the word "apple" with the apple. In this way, such an individual could start creating a communication language.

The human capability to teach, learn, and invent languages exceeds by far the communication capabilities, and even the intentions, of the most competent leader of any pack of wolves. Together with the enhanced thinking capability it provides, it may be responsible for our having climbed to the top of the food chain.

3.2 Is Our Brain "Intended" for Science?

This question comes up often in popular philosophical discussion, usually raised by someone who answers it in the negative. My answer happens to be positive.

The word "intended" appears between quotation marks to emphasize the fact that natural traits are not really intended for anything. The statement that a given trait was "intended" to make this or that ability possible is shorthand for the claim that this trait developed by accident, but natural selection favored the person who was blessed with it by virtue of that ability.

Everything we learn about the world on our own, as individuals, is the result of lessons learned from experiences. This is outright science, and every baby is a born scientist. The utility of each lesson learned depends on the number of situations in which we can apply it. This is why our brains excel in identifying the patterns to which each special situation belongs. The more patterns an animal is able to recognize at any given moment, the greater its chances are of surviving this moment, or of taking advantage of the opportunity this moment presents for surviving in the future. This makes looking for patterns in advance a favorable trait.

Because of this, our brain is built to love the process of pattern recognition; it likes to recognize patterns, and it keeps investigating because of the pleasure this causes, not because it is useful (just as we and other animals have sex because of the pleasure it causes). I expand on this point, among others, in "3.4 Meta-beauty."

This explains why our brain is explicitly "intended" to engage in all the activities involved in science, and why scientists enjoy their work. What we call the "scientific method" merely implements at the level of society what we already implement at the individual level.

Naturally, the scientific method was improved by its transformation into a social tool, both because of the sharing of information (the wise learn from the experience of others), and because various principles that individuals acted on intuitively were formally articulated, reducing the likelihood that someone would forget to apply them. The capacity to cooperate with others and to be able to describe things formally are abilities of the brain selected by natural selection for the benefits they bring. As already mentioned, there is no agenda behind the way the brain works. It works in a way that benefited its owner in the process of natural selection, and it so happens that continual research, for its own sake, is a successful survival strategy.

Animals also exhibit a certain level of such research, and one of the words invented to describe the urge to perform it is "curiosity." Because curiosity is also a useful trait, human society has learned to elevate it above the individual level, and advanced countries fund basic research, the results of which are not known in advance. Thus, there is no component of the scientific activity that our brain is not "intended" for, and the scientific activity of society is intended (without quotation marks) to grant society as a whole the advantages that the ability to engage in scientific activity grants individuals.

Note, however, that although our brains have been selected to investigate, not all of us excel at it. The reasons for this are many, and certainly the combination of natural talent and the tendency to choose the profession we excel at is one of them, but I think that there is a more important reason, which is also the reason why some claim that our brain was not "intended" to engage in science: engaging in science is an effort.

In his book, *Thinking, fast and slow*, Daniel Kahneman, describes the brain as a combination of two subsystems. System 1 is unconscious, agile, energy-efficient, intuitive, and imprecise, designed to enable us to make quick decisions because in many situations we don't have the time to analyze in an orderly way the possible courses of action before choosing the best. System 2 is conscious, precise, analytical, slow, and much less energy-efficient, which we use when we have all the time in the world and it is important to make as accurate a decision as possible. In addition to describing the characteristics of the two systems, Kahneman tells us three things about them. Different people use the two systems to different extents. Some rely almost exclusively on the first system, others use the second more frequently. The use of the first system is a given. We cannot avoid it. The use of the second system is the result of a decision. Scientists, whether naturally or as a result of scientific education they received, tend to activate the second system more than others.[67]

3.3 Will We Ever Know It All?

The scientific enterprise as a whole seems to be moving in one direction: the unification of all scientific theories into one theory that clarifies everything.

We discuss here the possibility of the existence of the longed-for theory,[68] not the question whether we will ever know the location of every elementary particle in the universe.

---

[67] This is especially true of times when the scientists deal with science. In everyday life, they sometimes let go.

[68] A unified theory is necessary for "knowing everything" because if there are separate theories that contradict each other, it's clear that we don't know everything, and if there are several theories that are not contradictory, they may be regarded as one grand unified theory.

This question, which many people have dealt with, is of special importance to me, as can be gleaned from "4.1 The Meaning of Life." In "9.7 Beyond Doubt," I point out the importance I attach to asking questions for advancing our understanding of reality. But what will happen after we understand everything?

In my opinion, people who find the meaning of life in trying to understand the universe can sleep without any care in the world, knowing that there will always be things that we will not understand. For the same reason, the "God of the gaps" can continue to sleep undisturbed in the knowledge that although his living space is continually diminishing with the accumulation of human knowledge, there will always be room for him to hide.

What is my conclusion based on? Certainly not on any prophetic ability. It is a conclusion that can be inferred from several facts that I believe are already known. I describe these facts below.

3.3.1 *Fact 1: Gödel's incompleteness theorem*

Roughly, this argument goes as follows: any consistent theory that is strong enough to describe natural numbers is incomplete, in the sense that in any model that conforms to it, there are real claims that cannot be proven or refuted based on the axioms that make up the theory.[69]

What does this say about science? It means that if and when any theory that presumes to describe all of reality is proposed, reality will be a model of this theory, and it will be possible to formulate claims about this model that will be impossible to prove or refute relying exclusively on this theory.[70] In other words, the theory will not really describe "everything," and there will still be questions we won't be able to answer based on it.

---

[69] Recall that a scientific theory is a collection of axioms together with the conclusions following from them.

[70] Note that although Gödel's statement discusses theories that allow the representation of natural numbers, it does not deal with natural numbers only. Because we see natural numbers as part of reality, any theory that describes reality should also describe them.

3.3.2 *Fact 2: The essence of an explanation*

When we try to explain something, we always base our explanation on things we accept as true. If we are asked about the reason why these things are true, we may try to explain them by means of other things we believe to be true. If we continue doing this *indefinitely*, we will eventually be left with a collection of claims that we believe are true, but with no way of basing our belief on any explanation.[71] The question, "What is the explanation of the basic claims that we accept as true?" will therefore remain forever relevant.

3.3.3 *Fact 3: Our reliance on qualia*[72]

In "1.3 Qualia: The Limits of Understanding," I tried to explain why *qualia* (the way we experience our sensory input) will forever remain unexplained. Contrary to the clauses dealing with the two previous facts, the conclusion of the present clause is not particularly encouraging. In the previous clauses we have seen that there will always be questions whose solutions are worth pursuing, but the present clause makes it clear that there will be questions that are not worth investigating because we know in advance that we will never know the answers to them.

3.4 Meta-beauty

The connection between the acceptance of a scientific theory and its aesthetics has been the topic of extensive discussion. Some have expressed astonishment that a text that succeeds in describing nature objectively (in other words, a scientific theory) also meets the criteria of aesthetics, which are usually associated with a subjective set of considerations.

---

[71] We cannot base the explanation of these claims on claims that we explained through them because this will be begging the question.

[72] The facts described in the first two clauses are accepted by most scientists and philosophers, but the argument in this clause is, for the time being, my own conclusion.

The word *beauty* stands for a variety of pleasant feelings we encounter in different situations. We consider our ability to perceive beauty as an important feature in defining ourselves as human beings. It is not surprising, therefore, that this topic has been discussed at length in scholarly articles as well as in the mass media.

The way beauty is presented and the vantage point from which this is done naturally change according to the approach of the person addressing the topic. My inclination is to trace the evolutionary origins of human traits, and this is the angle from which I wish to shed light on the topic of beauty.

I do not deal here with the type of beauty that attracts us to the opposite sex. This type of beauty is clearly derived from evolutionary selection, and almost everyone who dealt with it approached it from this direction. The types of beauty I am dealing with here are the beauty of science and of art.

Should we be surprised by the aesthetics of science? The connection between the acceptance of a scientific theory and its aesthetics has been widely discussed, and some have expressed astonishment that a text that succeeds in describing nature objectively (a scientific theory) also meets the criteria of aesthetics, which are usually associated with a subjective set of considerations. This astonishment reminds me of the sense of "enlightenment" I had as a child, when I noticed that all the actions nature requires us to perform in order to survive are "miraculously" actions that cause us enjoyment. This appeared to me at the time like a kind of gift, nearly attesting to the existence of a loving God. Unfortunately (or luckily), it took me a very short time to realize that this phenomenon is in reality necessitated by evolution and doesn't require divine intervention: an animal that doesn't enjoy carrying out the activities that are essential for its survival will simply not survive.

The situation with beauty is similar. Consider the description of the elements of that miraculous aesthetics: symmetry, simplicity, unity. These are precisely the features that allow short and catchy text to describe a wide variety of phenomena. Should we be surprised that we feel a sense of beauty and transcendence when we discover these qualities in a theory? It is natural that evolution "directed" us to prefer explanations that correspond to known facts, that are easy to remember, and whose implications are easy to figure out. The sense of beauty that we feel in the face of a theory is suspiciously compatible with the degree of its instrumentalism, and I believe that even those who do not subscribe to instrumentalism will admit that the instrumental criterion is ultimately the only one available to us, even if we seek an explanation that is real and not merely useful.

One may wonder how evolution could have influenced our preferences in the field of science, which has existed for so short a time. After all, human consciousness hasn't appeared on Earth until recently. The answer is that our preferences in science are derived from a cognitive ability that has preceded human consciousness by hundreds of millions of years: the ability to identify patterns. Some level of this ability exists in all animals that have a nervous system, and it forms the basis for research on these animals' learning ability and memory. The formulation of scientific theories is only the generalization of this ability, its transfer into the realm of the conscious and the public.

Anyone who has experienced a sudden discovery, understood a scientific explanation, or even solved a mathematical or scientific puzzle, must remember the excitement these experiences produce. Excitement and emotions are usually attributed to deep layers of the brain, which existed long before consciousness. This experience reinforces the sense that the source of our scientific preferences lies in the natural selection of animals that enjoy identifying patterns and thus achieve higher survivability.

Summarizing the above, we can say that the beauty we find in science has more to do with us and the way we evolved than with nature. When we succeed in formulating a simple law that captures large aspects of nature, we find this formulation pleasing, elegant, or beautiful. Many aspects of the behavior of nature may have escaped such a formulation, and some may escape it forever. We ignore all of them when speaking about the beauty of science.

This is not the whole story, however. As you can see in "8.4 Divine Symmetry," nature also makes a nontrivial contribution to the feeling of beauty that science evokes in us.

3.4.1 *Aesthetics in art*

Evolution has imprinted on us the enjoyments that are vital to our survival. These enjoyments were instilled in us even before we could understand their importance for survival, and remained available even in situations in which there is no need for an act of survival proper. Similarly to other animals, we tend to pursue these enjoyments not necessarily for the end that evolution "had in mind." This explains why people keep eating tasty food beyond the amount needed to maintain their bodies, why they continue having sex when they can no longer conceive, and so on. My argument is that the mechanism behind a significant part of our enjoyment of art is similar.

What are the mental qualities on which the enjoyment of art depends? One of the most important ones is the enjoyment of identifying patterns. This is consistent with the contribution that certain explanations make to the enjoyment of the work of art. These explanations make it easier to identify the patterns of the work of art, or are themselves the context by which the work is associated with pattern identification. This also explains the tension and mystery we feel when we first begin listening to a musical piece, and the enjoyment we experience later. This sequence of events is remarkably similar to the one that describes grappling with a mathematical or scientific puzzle. An artist who is seeking a way to express a certain feeling through his art and all of a sudden finds it, often uses the phrase "suddenly everything fell into place." A scientist who all of a sudden understands another aspect of reality uses the exact same expression.

I don't believe that pattern recognition is the alpha and omega of the enjoyment of art. I claim only that it occupies an important and central place, and that, moreover, it is shared by all of us and has an evolutionary reason. I am aware from personal experience of aesthetic enjoyment that does not involve the identification of patterns, and I cannot necessarily situate such instances in some explanatory scheme. I deliberately used the expression "aesthetic enjoyment" rather than "artistic enjoyment" because not always is there an artist behind the cause of such enjoyment. One of the "landscapes" that I loved to watch as a child, especially when I lived in Europe, was that of cloverleaf interchanges, especially viewed diagonally from beneath. This perspective makes it possible to perceive both the rounded structure and the height at which this massive construction "floats" in the air. When I tried to trace the origins of this enjoyment, I finally found them with the physical power that these structures transmit.

It is easy to understand why evolution has "trained" us to enjoy being in the presence of physical power that is not threatening to us, but is potentially threatening to our foes. Quite a few works of art build on this type of enjoyment, and in my opinion, this is the source of the magic of some of the works of Israeli artist, Dani Karavan. I had an almost religious aesthetic experience when, after spending the night in a tent in the Sde Boker area, I opened my eyes in the morning and took in the landscape. I think that this experience was also related in some way to the impression made on me by the giant force needed to create such an enormous flat area bordered by mountains. I had a similar experience when I climbed the slopes of the Ramon Crater, and its fringes were suddenly revealed in the sunrise.

A slightly different type of enjoyment is that related to classical art. Perhaps this enjoyment stems from admiration for the artist's skill. There were times when the degree of realism of an image was a measure of its quality, and in my eyes this realism still evokes the same admiration, which may be nothing but the urge to resemble the object of admiration. I have a similar sense regarding surrealist art.

In sum, the experience of beauty is a manifestation of a set of things we are programmed to like because they serve our survival. Although this explanation does not account for the neurological mechanism that is responsible for the enjoyment of beauty, it is still useful, for the following reasons:

1. It can serve as a step toward an explanation of how the brain processes the enjoyment of beauty.
2. It may be the best explanation we can hope to reach because no single universal explanation, at the level of neurological mechanisms, may exist. This is because at the neural level, similar mechanisms may be implemented differently in different people, as is often the case in experiments in artificial neural networks, which orient themselves differently but in the end provide similar answers.

Further Reading on This Story

Ramachandran, V. S. and Rogers-Ramachandran, D. (2006). "The Neurology of Aesthetics." *Scientific American Mind*, October issue (Volume 17, Issue 5), pages 16–18.

Arthur J. M. (2002). *Einstein, Picasso: Space, Time, and the Beauty That Causes Havoc.*

3.5 They're Cute, but They're Not Human

Certain questions that bother people and occasionally arise in online discussions are the result of over-anthropomorphism.

Consider, for example, the following question that came up on Facebook at some point: Are there transgender animals?

Naturally, the questioner knew that animals don't engage in surgery, so it is clear that he wasn't asking whether there have been animals that underwent sex reassignment surgery. I also suppose that he was not referring to natural sequential hermaphroditism observed in some species. He was asking whether there are animals that identify in themselves a mismatch between their gender identity on the one hand, and their designated or biological gender on the other.

My answer to this question contains two parts.

The first part is that to reach a definite answer to the question, we need to have a chat with the animal and ask it what it feels, because we have no other way of learning about its feelings about the matter.

The second part is that precisely because we cannot ask the animal such a question, we can infer with relative certainty that the answer is negative. In other words, we can conclude that no animals identify in themselves a discrepancy between their gender identity and their designated or biological gender. The argument in the second part is based on the fact that in order to sense such a mismatch, the animal must be equipped with understanding of such terms as gender identity, gender designation, and biological gender. To feel this kind of discrepancy, the animal must be capable of conceptualization, which, to the best of our knowledge, exists only in humans in any real and useful way. The connection between the ability to conceptualize and language, which I pointed out in "3.1 Think Before You Talk," makes the connection between the two parts of the answer.

Thinking about the possibility of transgender animals is therefore the result of over-anthropomorphism, where our empathy makes us think that animals are more like us than they actually are. This is only one of a variety of questions with regard to which people tend to make the same mistake of over-anthropomorphism. For example, those who think that animals eat to survive are mistaken. Animals eat because they are hungry. They also prefer certain foods over others because these foods cause them enjoyment. Nor do animals have sex for the sake of the next generation or to ensure the survival of the species. They do so because they have an urge to do it and because they enjoy it. We carry out the same activities, also for the reasons cited above, but also because we, contrary to animals, understand the connection between the acts and their consequences for the longer term.[73]

Note, however, that homosexual animals do indeed exist. The reason for this is, naturally, the fact that homosexuality, like heterosexuality, stems from impulses and is not the result of an analysis of actions and their consequences. Therefore, only in human homosexuals is it possible to find behavior that does not match their sexual orientation, as humans they may conduct sexual intercourse with members of the opposite sex, not because they enjoy it but because they want to have offspring.

3.6 Philosophical Movement: Do Plants Feel Pain?

Once in a while, an article appears in the press about plants that grow better when a certain type of music is played for them, suffer from pain, or become frightened of people who have caused them pain in the past. I presume that most readers do not believe these claims, but I thought it would be instructive to discuss the reasons for disqualifying them.

---

[73] An extreme example of over-anthropomorphism is exhibited by the people who accept Doctor Dolittle-type claims like the one presented in "The incredible story of how leopard Diabolo became Spirit — Anna Breytenbach, 'animal communicator'."

Scientific articles dealing with the subject completely reject such claims,[74] generally based on two types of investigation.[75] The first type includes research that studied the structure of plants and found that there is no central nervous system capable of conveying information about pain, and no sensory organs that may trigger it. The second type includes "behavioral" studies, in which the claim is examined in practice, without pretending to know the mechanism behind it. The somewhat eccentric title of this story was meant to express the claim that both the sense of pain and the ability to learn are closely related to the ability to move, and neither pain nor learning could have developed without this ability.

This conclusion is highly probable, and no research is needed to prove it. It is enough to take evolutionary considerations into account. Ask yourself how are pain, feelings, and thinking related to the increased ability of organisms to survive. You will find that these abilities make no contribution to survival if the organism is unable to move, and that they are not byproducts of other abilities benefiting organisms that cannot move.[76]

Consider, for example, the function[77] of the experience of pain. Its "function" is to make us try to avoid the cause of pain.[78] Except for chronic pain caused by some malfunction, if the pain is caused by an external factor, avoiding it involves getting away from it, in other words, movement — escaping from fire, for example; if the source of the pain is

---

[74] As you can see if you read the article on plant perception in Wikipedia and follow the links that appear there.

[75] A more comprehensive presentation of research achievements in this field can be found in Prof. Daniel Haimowitz's lecture on Coursera.

[76] Some of our abilities, although not directly contributing to our survivability, are byproducts of such capabilities. Altruism, for example, could be such a trait, as I explain in "2.2 Why Do We Empathize?

[77] I tend to use teleological language, which ascribes purpose and intention to things, but I do so only because it makes for clear and concise expression. In reality, things happen without a guiding intention, and the rationale presented as their motive represents the fact that, in retrospect, of all things that have evolved by accident, those that were selected by evolution have been given prioritybecause of this rationale. To make sure that the teleological formulation does not mislead, I occasionally use quotation marks to emphasize the fact that this is only a manner of speaking.

[78] Pain is not required for the healing process itself, which is why we allow ourselves to use anesthesia and pain killers.

internal, avoidance of pain may involve the cessation of movement — for example, resting and letting the injury heal. When the organism is unable to move, the experience of pain has no function and therefore it is highly unlikely that the ability to feel pain will develop in it. Note that I am talking here about the experience of pain rather than an automatic response to something that might be harmful. A local automatic response does not require an experience that can be referred to as "feeling by the organism" because it can be based on local sensing.[79]

To understand what I mean by the term "pain," remember that feelings, emotions, and impulses are what underlie the will. Will, in the sense in which we perceive it, is useful for carrying out the coordinated action of different parts of the organism. It is not necessary for producing a local response of a part of the organism, but for example, when the organism tries to escape, the movement of both legs must be coordinated. The benefit of will is limited to situations in which it is possible to choose between different activities. All parts of the stone fall in a coordinated manner, but will would be of no help to it because it has no other choice.

What is true for pain is also true for many other bodily sensations: a sense of excess heat or cold is "intended" to encourage us to look for a place where the temperature is more comfortable, which involves movement. Itching is intended to encourage us to act to remove the source of the itch. Smells are intended to encourage us to eat the source of the smell, if it is appetizing, or to get away from it, if it elicits disgust. Naturally, these are merely examples. Smells are also related to choosing mates, avoiding predators, and so on. All these actions involve coordinated movement of different body parts in situations in which various options for moving are available.

The ability to create for ourselves a "picture of the world," to think about it, and to imagine scenarios that will develop as a result of actions we will take is relevant only if we can move. This is because, unlike other

---

[79] In the absence of appropriate terminology, I distinguish between the term "feeling/experience" and the term "sensing." To describe the relationship and the difference, I attribute to these words, it is possible to use the following equation: feeling = sensing + caring about it. A good example of the way local sensing is exhibited in plants can be found in an article that describes the way some plants react when in danger.

abilities involving responses limited to the body's chemical arsenal, the ability to move requires us to choose between different alternatives whose usefulness depends on the accuracy of the picture of the world that we have formed, and on the accuracy of our thinking in identifying the expected results. When we cannot move, and our ability to react is limited to a chemical reaction, forming a picture of the world and running simulations based on it have no advantage, and therefore it is highly unlikely that evolution would encourage the development of these capabilities.

What about the ability to enjoy art, for example, a lovely melody? This ability is a byproduct of our ability to think (see "3.4 Meta-beauty"), and therefore it is unlikely to be found in organisms that do not think — and we have already pointed out the connection between thought and movement.

Does this mean that the experiments conducted to test empirically whether plants have feelings were unnecessary? Not at all. In addition to the fact that these experiments teach us quite a bit about the functioning of the plants, those who agree with what I have said so far can see the results of these experiments as further confirmation of the theory of evolution, that is, the realization of predictions that can be made based on the theory.

Chapter 4

Being Human: Longing for Purpose

Seeking a purpose and meaning for life is probably unique to humans. It is not a direct product of evolution but its byproduct. In this chapter, we discover that a person can find meaning and purpose in life even if he doesn't commit to fulfilling the commandments of an imaginary entity.

4.1 The Meaning of Life

Religious people live to fulfill the commandments of God. What do atheists live for?

First, we should remember that all creatures live well without trying to have a purpose in life. At the lowest level, we also can live like this. Evolution has imprinted in us the will to live, just like in the rest of the animals, and the reason[80] for this is clear: creatures that do not want to live become extinct.

But humans are different in some ways from other animals. Our thinking ability has developed beyond that of any other species, and has created elaborate concepts. Among these are such concepts as "purpose" and "meaning."

---

[80] I said reason, not purpose. According to the scientific perception, things that happen in the world have a reason, but no purpose.

Evolution causes us to anthropomorphize many things. We routinely talk about ATMs eating our cards, and we often ask what computers and other sophisticated machines actually "want." Anyone who has seen mechanics try to fix a device or programmers debug software knows what I mean. It is only natural, therefore, that we would try to anthropomorphize existence as a whole, and expect it to aim for some purpose or to represent some meaning.

The ideas of purpose and meaning are entirely human concepts that do not represent any specific law of nature. And although we are creatures of nature, we are naturally looking for meaning. This is how nature created us. We appear to have no choice but to forever seek a meaning in life, above and beyond our natural urge to live — an urge that will not disappear, and thanks to which all of us who are not philosophers, and even most of those who are, will keep clinging to life.

Because nature has not endowed our lives with meaning, and because we ourselves created the problem of searching for meaning and we are the ones who suffer from it, we have no choice but to try to inject meaning into our lives. It is not meaning granted by the world or by God, but as long as it quiets our hunger for meaning, we are satisfied.

I can share with you a tale I'm telling myself. Meaning or no meaning, I am afraid to die. Because death is inevitable, I chose to dedicate much of my time to fighting the meaning of death. And, lo and behold, this war also provides a sense of meaning to my life. I conduct the battle through an honest attempt to understand how the world works.

There is a saying, which is quite characteristic of the New Age approach: "knowledge is the mother of boredom." This is a pessimistic view of reality. I see the same facts in a different way, which is much more optimistic. But first, I must share a joke about the difference between a pessimist and an optimist: the optimist thinks the world is the best it can be; the pessimist fears that this is exactly the case.

But what are those facts that I view optimistically? Our lives, by and large, are a collection of experiences. Some of these experiences repeat themselves, and excite us each time afresh. Once we understand completely the physical mechanism behind the experience, however, something of its flavor is lost to us — the mystery has gone out of it. This is the reason for the expression "knowledge is the mother of boredom."

But this expression does not take into account the supreme joy in the sheer fact of understanding and knowing. Perhaps some of the flavor of the experience is lost, but it is replaced by something much better. Indeed, once I have completely understood the mechanism behind a certain experience, I no longer need to experience it again and again to enjoy it. In many ways, it has become part of me. Put differently, with respect to this experience, the dimension of time has collapsed into one point, which is the moment of insight. What does it mean that I no longer need to repeat the experience over and over to derive full pleasure out of it? It means that as far as this experience is concerned, I have lost the need for eternal life because I have already extracted from it all that I could hope to derive from it in eternal life. With each area of life that I can bring to this situation in which I understand it completely, my need for immortality diminishes, and I move one step closer to finding a cure for the disturbing problem of the finality of my life.[81]

4.2 Other People's Lives

About altruism as the meaning of life.

In "the meaning of life", I described the meaning I found for my life. I stressed the fact that this was a personal choice. In the present story, I would like to aim the spotlight on the choice of people who dedicate their lives to the benefit of others.

It is reasonable to argue that if I dedicate my life to the benefit of other people's lives (OPL), I must first find a meaning in the lives of those others. This argument may be answered as follows: This is not meant to be an absolute answer to the question, "What is the meaning of life?", but rather an answer to our emotional needs to find meaning in our lives.

I believe that there is no external factor that gives an objective meaning to our lives or requires that such a meaning exist. The need for such meaning is the result of our psychology: animals and people who are unable to devote time to reflection do not bother themselves with this question. Therefore, the answer to it is not absolute truth, and each person

---

[81] As explained in "1.3 *Qualia*: The Limits of Understanding," feelings we don't want to lose, like love, are immune to this pursuit of understanding.

must discover it individually. If concern about the wellbeing of others is the answer for some to their need for meaning — good and well. In any case, finding the meaning of OPL should be done by those others and not by those who have chosen to dedicate their lives to the welfare of others.

One name that many mention as an example of a person who chose OPL as the reason for her life is that of Mother Teresa, an Albanian Catholic missionary nun, humanitarian activist, and founder of The Order of the Missionaries of Charity, which is claimed to have helped tens of thousands of people. Her work, humanitarian aid, and support for the poor of Calcutta have made her one of the most famous women in the world, and a symbol of righteousness and integrity in the eyes of many.

Note that I'm not delving into the actual motivation behind the activity of Mother Teresa. Many people claim that her motivation was far from altruistic. I only use her symbol to clarify what I mean when I talk about a person who chose OPL as the meaning for her own life.

I will focus on two other examples and share with you the insights I have derived from them.

The first example is that of George Price, a highly talented eccentric man who made a significant contribution to many fields of knowledge and never managed to translate his insights into success in real life — specifically, money or prestige. He spent most of his life in the US, but after a serious illness made him aware of the finality of his life, he decided to leave everything, including his family, and move to Britain to try to kindle a scientific revolution that would bring him world fame.

Price intended to base his revolution on his solution to the puzzle of altruism, a behavior that contributes to the general welfare while diminishing the chances of reproduction of the individual who practices it. The result of his work on the subject was the Price equation, which many consider to be the best solution that evolution can give to the phenomenon of altruism. This equation explains how, in some cases, altruistic behavior can increase the distribution of the genes that cause it, although the specific individual who practices it may be harmed, by increasing the fitness of the group to which that individual belongs. This is the idea of group selection, which is still highly controversial among evolutionists.

But Price himself was troubled by another aspect of the explanation he found. He realized that his solution was a representation of altruism as egotism in disguise, which deeply frustrated him. This frustration had a profound effect on the rest of his life and its end. By rebelling against the conclusions that follow from his equation, he became an extreme altruist and gave away all his possessions to help homeless people he didn't know. He sought to prove with his body that non-profit altruism existed after all, and he did. The result was that he himself became a complete pauper and committed suicide. An instructive description of Price's life and work can be found in Oren Harman's book *The Price of Altruism*.

The personal insight I gained from Price's life story is that if he had read my explanation of altruism in "2.2 Why Do We Empathize?," perhaps he would have spared himself the frustration that led to his suicide, because this explanation, contrary to what one might expect from an evolutionary explanation, describes pure altruism not intended for profit as a necessary byproduct of our other features.

The second example is that of my father. In her last years, my mother suffered from a debilitating muscular dystrophy that more or less confined her to an adjustable armchair, in which she had to spend the last three years of her life. As her condition deteriorated, even before she was completely confined to that armchair, my father devoted an increasing amount of his depleted physical and mental resources to her welfare and wellbeing, until these became the center and purpose of his life. After a life characterized by constant activity — work, hobbies, and ceaseless learning — he abandoned everything and made my mother's welfare his only goal. This goal filled his life and helped preserve his vibrant creativity when it was necessary to find creative solutions to my mother's increasing physical difficulties. Although the description of these solutions is interesting in itself, it would deviate from the subject of this story and therefore I do not proceed with it.

At the end of December 2012, my mother passed away, and after we recovered from the shock, it became clear to my father's immediate family that his life became completely emptied of all content, and that he had difficulty rehabilitating himself. We are trying to bring him back to the activities he used to do before he enslaved himself to my mother's illness, but his physical and mental deterioration make it very difficult. Even if these efforts succeed (and results so far have been encouraging), I must still draw the following conclusion: If OPL become the only meaning of your own life, the meaning of your life depends on your ability to help others. Therefore, if they die, or if you lose the ability to help them, there is a danger that you will find the rest of your life meaningless.

Therefore, although the wellbeing of others is a good and worthy goal, it is desirable to find an additional meaning for our lives, one that will continue to serve our needs for a meaning for our lives, as long as we have the ability to think about such a need.

Chapter 5

The Selfplex Rules the World

Our tendency to identify the selfplex with the self has serious, and at times dangerous, implications

Humans differ from most animals in one important point: they hold a collection of thoughts and ideas (a memeplex) in their brains, in which they tend to recognize their "self" more than in their physical bodies. The name "selfplex" has been coined by Susan Blackmore to represent this collection of thoughts and ideas.

The identification of the selfplex with the self is behind many phenomena, some of which are listed below:

1. The ability to take comfort in the mistaken notion that life goes on after the death of the body.
2. The thought that replicating our mind in a computer will enable us to continue living after the death of our body.
3. The thought that one's quantum teleportation (which I don't believe will ever be achieved) is a means of transportation rather than a way of destroying oneself and building a replica somewhere else.
4. The willingness to sacrifice our lives to promote an idea.
5. Asceticism, where people rebel against the demands of the body, and even renounce the opportunity to pass on their genes.
6. The custom of commemorating people through a contribution to the goals they tried to achieve in their lives.
7. Cognitive dissonance: Normally, when two ideas contradict each other, we tend to discard at least one of them. This is difficult to do when

these ideas are part of the selfplex. Cognitive dissonance is the result of inconsistencies in the selfplex.

8. And most important, the fact that in addition to their propensity to procreate, which is an impulse imprinted on them by evolution so that their genes continue to exist, most people also seek to produce students (spiritual offspring) so that their ideas continue to exist. Writing this book is one of the ways in which I personally succumb to this tendency.

The stories in this chapter illustrate some of the roles the selfplex plays in our society and introduce a new term, the grouplex to describe the collective part of the selfplexes of a group of people.

5.1 Are the Religious Capable of Faith?

Cognitive dissonance in religion.

Debates with religious people have taught me that the answer to the question in the title is not obvious. Until these discussions, I used to think that we all use the expression "I believe X" to express a certain level of probability that we attribute to the truth of argument X. That level may not stand at 100%, but it is high. These discussions, however, have taught me that when a religious person uses this verb in relation to his religious faith, in many cases, even if he is not aware of it, he means something entirely different. My argument is based on the fact that it often seems that the religious person knows what is true, and yet he claims to believe in something different.

I already mentioned several examples of this in the story "2.1 About Morality," in which I pointed out that religious people who try to persuade others that the source of morality is in religion, always base their claim on commandments such as "You shall not murder" (Ex. 20:13), and never on commandments such as "Anyone who desecrates the Sabbath is to be put to death" (Ex. 31:14).

This phenomenon stems from the fact that in their heart's heart it is clear to them that "You shall not murder" can credibly express a moral imperative, whereas "Anyone who desecrates the Sabbath is to be put to death" contradicts their own inner moral imperative.

This is not an isolated phenomenon, but something that recurs frequently. Particularly instructive are the examples in which an atheist quotes explicit religious commandments, without expressing an opinion about them, and in response receives personal attacks accusing him of a negative attitude toward religion or of gratuitous hatred of the religious. One of the discussions I carried out on Facebook serves as a good example.[82]

In this discussion, I published a few factual claims on the topic referred to in Jewish religion as *chalitza,*[83] and I received reactions such as: "You are busy spreading hatred and adding fuel to the fire... We understand that you are opposed to religion, we understand that you think that every believer is stupid, and we understand that you are fighting against the spread of Orthodoxy, as you describe yourself in your profile. The argument in the village is about building a synagogue, about its size, and at times, also about Jewish identity in the State of Israel. And instead of trying to advance the discussion in a productive way, you are busy pushing your hateful agenda by sharing provocative and shocking articles. Release us from your demagoguery, and stop jumping around here like a pyromaniac with a lighter."

This phenomenon is interesting, encouraging, and disappointing at the same time.

It is interesting because, as stated, my posts included factual statements and not judgmental ones, so what was it that made the connection to that opinion in the mind of the responders? Apparently it came from themselves, because it is not stated in the post. In other words, they themselves understood that, based on these facts, religion looks bad, and because they believed that these facts required those who were aware of them to think bad things about religion, they concluded that mentioning these facts attests to my attitude toward religion. I find this interesting psychologically.

---

[82] Living in Israel, I had such discussions only regarding the Jewish religion but I'm quite sure people encounter similar phenomena with other religions as well.

[83] According to the Jewish religion, the widow of a childless man must marry his brother. Chalitzah is the ceremony the widow and the brother must perform to avoid this duty. The point I was emphasizing is that a deaf woman is not allowed to perform this ceremony, and is therefore forced to marry her husband's brother, even if he is already married.

It is encouraging because the facts seem to speak for themselves, and even religious brainwashing cannot prevent people from drawing conclusions about the nature of religion.

It is disappointing because the psychological exercise they have pulled on themselves worked. They attributed the opinion that had formed in the logical part of their brain to me. In doing so, they prevented the brainwashed part of their brains, which eventually determines their behavior, from adopting the conclusion they themselves had reached.

How is it possible that a person who understands that a particular religious commandment is immoral, and that he himself would never observe it, continues to claim that this commandment was handed down to us by God Almighty, who knows everything and wishes our best? Such a person must be in a state of cognitive dissonance.

It seems to me that the unwillingness to continue in this state of dissonance is the main source of a phenomenon called *apologetics*. Whenever a religious person is confronted with a religious claim, whether moral or factual, that contradicts his knowledge, he tries to "soften" the contradiction by claiming that without a convoluted interpretation, we cannot understand exactly what is written in the Scriptures. For some reason, it is easier for him to add to his set of beliefs the belief that what is written in the Torah is not what God intended us to understand, than to give up his faith in the truthfulness of the things written in the Torah.

I'm pretty sure that many religious people who claim that such apologetics satisfy them do not really feel this way. The philosopher Daniel Dennett probably came to the same conclusion, which he described in a lecture called "Good Reasons for 'Believing' in God," Dan Dennett, AAI 2007.[84]

5.2 Cognitive Dissonance Outside Faith

Hostility towards vegans.

Some people clearly feel a cognitive dissonance between their natural tendency to empathize with animals and their habit of eating them. This is well demonstrated in an ingeniously orchestrated Brazilian prank.[85]

---

[84] Good Reasons for "Believing" in God, Dan Dennett, AAI 2007.
[85] Brazilian prank.

In this prank, an actor plays the role of a supermarket butcher and invites people to taste, and then buy, a new type of sausages he prepares on the spot using a special sausage making device. The device is just a box with a handle he turns to prepare the sausages and an opening where the sausages get out.

The customer does not know that the box is open at its bottom and a person is hiding underneath, just pushing readymade sausages out of the box.

The customer tastes the sausages and after he agrees to buy some more, the actor acts as if the meat supply in the machine is out and tells the customer to wait a second and let him refill the meat supply. Up to that moment all goes well and the customer is happy but then, instead of putting readymade meat in the machine, the actor puts in a living pig that is taken aside by the person hiding under the machine, and immediately starts turning the handle getting new readymade sausages from the person below.

The customers are horrified, of course, and refuse to take the sausages since they can no longer repress the knowledge they had all along that pigs are killed in the preparation of the meat that is used to make the sausages.

Next time you encounter a person attacking vegans telling them that they are supercilious, remember that it's just their own feeling of guilt that they are expressing.

There are many examples of cognitive dissonance both inside and outside religion[86] but there is a crucial distinction: a non-believer that becomes aware of a cognitive dissonance can readily correct it,[87] while a religious person that becomes aware of a cognitive dissonance that stems from religion, faced with the choice between the claims or commandments of religion and his own feelings about truth or morality will usually resort to some convoluted apologetics.

---

[86] You can find many of them here.

[87] The fact that he can correct it doesn't mean that he does. In many cases he just suppresses it.

5.3 Sense and Sensitivity

The distinction between feelings and reason often serves as a basis for critical arguments against the way secularists treat religious beliefs and religious people. These arguments are raised only because religious people include their religious beliefs in the selfplex.

Here are two examples of such arguments:

1. It is pointless to try to persuade religious people of their mistake using rational arguments because their belief is based on feeling rather than reason.
2. It is wrong to "hurt one's religions feelings" (an argument enshrined in the laws of many countries).

In my opinion, these arguments are not justified.

<u>In most cases, religious belief and religiosity are the result of indoctrination.</u>[88]

The majority of people who believe in God[89] do so because they were raised to believe when they were children. Is it possible to describe the education a person received as a "feeling"? Of course not! This education is nothing but indoctrination that takes advantage of children's tendency to believe what adults say. Children's inclination to believe adults has an evolutionary basis: it is "designed" to spare children the suffering and risk

---

[88] Religious beliefs and religiosity are also influenced by genetics, but indoctrination plays an important role in religiosity even for people who are genetically inclined toward religion. Changing genetics is difficult, but changing the influence of the environment is something both educators and students can achieve. Examples of such influence are easily discerned at the extremes: <u>according to the Global Index of Religion and Atheism 2012,</u> 47% of the Chinese are convinced atheists and another 30% are not religious. Only 14% of the Chinese are religious. In Ghana, by contrast, 0% are convinced atheists, 2% are not religious and 96% are religious.

[89] The word "majority" is important. Some people believe in God as a result of a decision to accept religion. The reason for this decision can indeed be described as emotional. For some of these people, it may be a desire to break free from responsibility for their fate and actions.

involved in learning from experience. One of the byproducts of this evolutionary trait is the ability adults have to indoctrinate children with any items of information they may provide, including erroneous ones that contribute nothing to the children's chances of survival.

Children believe what adults say not as a result of feelings but as a result of logic![90] It makes sense that the adult's knowledge of the world and of the ways of dealing with various difficulties is greater than that of the child. Therefore, from the child's point of view, it makes sense to believe what adults say.

Adults do indeed indoctrinate children, but I don't mean to attribute evil intentions to the adults doing so, because they indoctrinate children with ideas that they themselves believe to be true. Adults believe that they impart knowledge that will serve the children later in life.

Naturally, there are other periods in the lives of some people, beyond childhood, when they may fall victim to indoctrination. For example, people who have experienced disaster or loss may seek relief by turning to those who claim to have a cure for any problem. Such periods can be described as temporary relapses into childhood, when independent thought is neutralized and the words of the "spiritual teachers" are perceived as absolute truth.

Mechanisms of indoctrination explain how faith passes from person to person, but a full explanation of the phenomenon of religious belief also requires an understanding of the process by which faith was formed in the first place. After all, there had to be someone who started believing without receiving his faith from others.[91] What made this person start to believe?

Most of the time, the formation of faith is not based on emotion either.

To gain control over our lives, we try to identify patterns of cause and effect. This understanding has two roles: to gain control by triggering the

---

[90] I am not referring here to a logical inference on the part of the child, but to the logic that evolution imprinted on children's behavior.

[91] Note that I'm talking about belief in the supernatural and not about religion. Religion may have been, at least in some cases, a manipulation from the very beginning, but even in these cases it relied on existing beliefs, using them to justify the rules.

cause so that it would produce the effect, and to predict the effect that will be caused in the future by current circumstances identified as the cause of that effect.

When we fail to find a "natural" cause for a phenomenon, we tend to conclude that it was caused by someone who wanted to achieve it. In most cases this is a correct conclusion (the cheese didn't move by itself, so someone had to move it). In other cases, however, the conclusion is wrong because we were caught in the logical fallacy of the argument from ignorance: we found no natural cause and therefore assumed (sometimes mistakenly) that no natural cause exists. In such cases, when we fail to understand who it was that caused the effect, we tend to invent God.[92] This is no other than the familiar "God of the gaps." It is quite clear that there can be no other explanation for these gods, because when we understand exactly how a certain thing happens, we have no reason to attribute its occurrence to God. Only when we don't know how it happens do we invent imaginary reasons for it.[93]

We should not infer from the above that faith in itself is logical. My argument is that only the process that leads to its adoption is one that it is usually logical to use, and that faith is achieved because of exceptions to this rule.

If we don't want to follow the path of indoctrination, the only way we can honestly persuade is by reason. All human beings can reason, including religious ones, because it is the result of evolution. It is true that the feelings of a debater may prevent its activation, but if we don't use logical arguments, we are assured in advance that logic will not be activated and quite likely, we will not be able to affect another's opinion.

---

[92] The following is an example of hundreds of articles that deal with this topic. It is important to understand that the term "God" does not necessarily represent the God of the Abrahamic religions, but entities (animals, trees, inanimate objects, and even imaginary entities) to whom various intentions and capabilities that they do not have are attributed.

[93] Invent or accept: sometimes we don't invent anything in person, but a religious preacher comes along and offers us an imaginary explanation of what is happening, and in the absence of another explanation we accept the offer.

One of the important reasons for using logical arguments is that some religious people (for whom, as noted, logic is also important) try to present their faith as a result of logical considerations. One of the purposes of logical debate with them is to expose them to the understanding that their "logical" arguments are mistaken and are nothing but an attempt to retrospectively justify the conclusions they have reached in advance. Much of what I wrote on the blog and in the book was for just this purpose.

The previous paragraph leads us to the second argument.

<u>There is no such thing as harm to religious feelings</u>!

Let's start with the self-evident: religion has no feelings. Emotions are the domain of living beings and not of ideologies. Therefore, when we talk about "religious feelings," we are referring to the feelings of those who believe in religion. These are emotions that believers feel when someone denies, refutes, or ridicules their religious beliefs, but these feelings do not stem from the content of the belief but from the fact that someone denies, refutes, or ridicules the memes they identify with their self, what is commonly referred to as their selfplex.

Feelings of this kind do not characterize religious people only: every person feels offended when others try to undermine belief in what they regard as their most hallowed convictions. The difference between religious believers and atheists lies in the fact that atheists understand that for the sake of protecting their ideas, they can resort to logic, whereas the religious, whose belief is not based on reason, when they recognize this, turn to emotional blackmail.

In sum, religious believers do not believe because of their feelings, but their feelings are vulnerable because of their religious beliefs.

5.4 Religion Contributes to the Survival... of Itself Only

It is often argued that the success of the religions attests to their contribution to adaptation of humans to life in their environment. It is debatable whether religion ever contributed to the survival of

humans, but its survival is easily explained by its contribution to its own survival.

Religion is a collection of beliefs, commandments, rites, and rituals that together form a memeplex. The commandments in this memeplex describe rules of conduct that are not determined by legislatures, but are alleged to originate from a source external to humanity (generally, God, but also other factors, as, for example, the principle of white supremacy). The destructive power of religion derives from two sources: the special nature of this memeplex, and the tendency of children to believe what adults say.

All religions are a collection of memes (memeplex). A religion that survives is a collection of memes that preserves itself. I emphasize that it preserves itself and not necessarily the person who carries it in his mind. It does so, for example, by such memes as martyrdom for refusing to renounce one's religion, murder for the greater glory of God, conversion, or by all kinds of falsehoods about the scientific truths hidden in the Torah.

Why do I claim that it is a collection that preserves itself rather than its hosts? After all, martyrdom does not preserve the host. But if the meme of martyrdom for refusing to renounce one's religion had not been coined, all its hosts would have converted when forced to do so, saving their lives but forsaking their religion. Therefore, religion prefers to occasionally sacrifice a few people (who, from its point of view, are only a platform) for the sake of its own survival. Many Jews lost their lives during the Spanish Inquisition because of this meme, although others (crypto-Jews), publicly professed Christianity while secretly adhering to Judaism. This is what has happened throughout Jewish history, and therefore I claim that Jewish religion survived at the expense of its people. Under Shia Islam, life-saving fake conversions have become formalized (as part of what's known as *taqiya*).

But religion is not merely a self-sustaining memeplex. It is a memeplex that becomes part of the selfplex, hijacking the tendency of people to protect memes that they view as constituents of their self-definition. Moreover, by relying on memes that encourage the separation of the group of believers from the rest of humanity, religion becomes a grouplex and can further enhance its hold on people through memes that encourage the members of the group to prevent other members from going astray.[94] Religion derives a large part of its influence from this fact.

Religion is particularly dangerous because of a particular characteristic of human beings that is well described in Arthur Koestler's book, The Ghost in the Machine. Koestler laments the tendency we have as human beings to perform, as part of a herd, horrific actions that we would never dare perform as individuals.

Here we find the connection with the claim, often voiced, that many of the crimes against humanity were perpetrated by "secular" forces such as Nazism and Bolshevism. But although they are presented as secular phenomena, Bolshevism and Nazism are in practice, examples of religions, in the sense that they represent a collection of memes that are accepted uncritically, a collection that preserves itself without compromise. Bolshevism, Nazism and several other political ideologies are religions for all intents and purposes, even if they are not based on God, because they rely on axioms alleged to originate from a source external to humanity, which result in insane behavior. Similarly, nationalist fanaticism is a religion.[95]

Whether or not you agree that it is correct to call these phenomena religions, it is important to realize that all the evil in these phenomena stems from the same sources from which the evil in religions originates: a collection of memes, some of them wrong, that preserves itself at the expense of human beings. In some religions, there are memes or commandments that contribute to the reproduction of the human beings who hold them, but here, too, humans are used merely as a tool for the survival of the memes.

It is appropriate to call such a collection of memes a "thought virus," because, just like a virus, it can cause an epidemic and take control of all human beings. Although it benefits religion, it is most harmful to humans,

---

[94] The fact that religions include such memes casts considerable doubt on the assumption that religiosity has been encouraged by evolution. Normally, there is no need to force people to comply with their evolutionary heritage.

[95] Indeed, any ideology in which human claims are raised to the level of objective/scientific facts is a religion. A normal vegan knows that he is enacting his own preferences. He becomes religious if he thinks it is a global imperative and begins attacking butcher shops.

it kills many people, and if it eventually manages to take over the minds of all the people whom it did not kill, these will cease to be autonomous human beings.

The tendency of youngsters to believe adults exists among all animals that are capable of learning, and it is essential for the youngsters' ability to quickly learn what adults know, without being exposed to the dangers of having to discover this knowledge by themselves. A young monkey learns from the adult monkeys to be afraid of snakes, which saves him from the danger of learning on his own flesh the damage a snake can inflict. A situation in which adults know a lot more about the world than those who were just born is natural. This situation creates an evolutionary advantage for the newly born to receive the knowledge of their parents literally, whether it is transmitted by personal example or in words, without criticizing it too much. Youngsters who do so increase their chances of survival by intuitively applying Bismarck's quip: "Only a fool learns from his own mistakes. The wise man learns from the mistakes of others." It is not surprising therefore that the vast majority of people have the same religion as their parents.

At a certain stage, the knowledge that one's parents can inculcate is exhausted, and faith must be replaced by critical thought. Critical thinking, which begins at a certain age, also has an evolutionary advantage. The phenomenon of "youthful rebellion" is therefore natural, and was "intended" to initiate this type of behavior. The problem is that criticism is usually applied only to new knowledge that is currently being evaluated, and not to beliefs learned in early childhood. This is why there are also religious adults.

5.5 Selfplex and Grouplex Outside Religion and Ideology

In principle, the scientific method discourages the inclusion of scientific ideas in the selfplex but as a human endeavor, science is not always free of such inclusions.

The scientific method promotes doubt and evidence based search for truth. This approach prevents memes from becoming part of the selfplex since the scientist must be prepared to discard any assumption he made as soon as it is contradicted by evidence. But scientists are human and as such they often fall in love with some ideas (mostly their own) and integrate them into their selfplex to such a degree that they are willing to protect them even in the face of contradicting evidence. Grouplexes too, are encountered in science and Pythagoreans are sometimes accused of having drowned Hippasus in order to hide his discovery of irrational numbers.[96]

Selfplexes in science motivated Planck to formulate Planck's principle stating that "A new scientific truth does not triumph by convincing its opponents and making them see the light, but rather because its opponents eventually die and a new generation grows up that is familiar with it... An important scientific innovation rarely makes its way by gradually winning over and converting its opponents: it rarely happens that Saul becomes Paul. What does happen is that its opponents gradually die out, and that the growing generation is familiarized with the ideas from the beginning: another instance of the fact that the future lies with the youth."[97]

In his book, *The Structure of Scientific Revolutions*, Kuhn argues that besides objective logic and facts, scientific education also includes an element of indoctrination that directs scientists to commit to an existing paradigm, thus defining the questions of interest and the type of acceptable solutions. Kuhn's paradigms are examples of grouplexes, and the emotional and behavioral loads associated with grouplexes sometimes impede scientific advancement.

A recent example of the impediment caused by selflexes and grouplexes is the very long period between Dan Shechtman's discovery of quasicrystals in 1982 and the universal acceptance of his findings that earned him the Nobel prize in chemistry in 2011. During this period he had to endure harsh opposition from appreciated scientists. Linus Pauling, who got the Nobel Prize in chemistry in 1954, never accepted Shechtman's findings and expressed his hostility towards him saying that "There is no such thing as quasicrystals, only quasi-scientists."

---

[96] The discovery of the first irrational number.
[97] Max Planck, Scientific autobiography, 1950, pp. 33, 97.

The head of Shechtman's research group asked him to leave for "bringing disgrace" on the team.

The scientific endeavor discourages the establishment of selfplexes and the formation of grouplexes by encouraging competition between scientists. Using the scientific method that regards experimental findings as the final arbiter, it is built to ultimately overcome selfplexes that have been established and grouplexes that have been formed.

It is worth noting that by harnessing the battle between selfpexes to overcome stagnation, the scientific approach may weaken science based communities in their struggle with faith based communities that find it easier to unite under a charismatic leader or otherwise form grouplexes. This is one of the reasons why separation of church and state is so important.

Chapter 6

Evolution vis-à-vis Creation

The theory of evolution challenges the religious story of creation. Therefore, it is not surprising that as soon as the general public got wind of this theory, opposition to it began. Throughout history, opposition to evolutionary theory has worn many faces, but all its versions originated in religious belief. Contrary to what people sometimes think, the proponents of evolution have never claimed that intelligent creation is impossible, but that it did not happen in practice. Some supporters of evolution even tried to create life in a lab, and an example of it can be seen in Craig Venter's Synthetic Life project.[98]

6.1 The Unquestionable Fitness of Evolution

One of the arguments raised by opponents of evolution is that it is merely an unproven theory. Beyond the foolishness of the requirement that a scientific theory be proven, when the very criterion for the scientificity of a theory is that it is, at least in principle, refutable (whereas a proof, as is well known, cannot be refuted), the assault on evolution in this matter is particularly preposterous because of all the scientific theories, evolution is closest to being "proven."

The basis of modern evolutionary theory is a collection of mathematical theorems claiming that when certain conditions are met, evolution

---

[98] Craig Venter's Synthetic Life project.

occurs. The conditions in question are, largely, the existence of entities that can replicate with small changes, enhanced by their ability to succeed in coping with resources that are in short supply. The evolution that occurs in these conditions is made possible by the deaths or minimal replication of individuals whose characteristics do not allow them to compete for resources, against other individuals who are more successful in it. This is a purely statistical phenomenon, and there are quite a few mathematical theorems that describe these and other aspects of it.

One of the first theorems formulated in this field is Fisher's fundamental theorem, which states: "The rate of increase in the mean fitness of an organism ascribable to natural selection acting through changes in gene frequencies[99] is exactly equal to its genetic variance in fitness."[100] A historical fact that evokes amusing associations is that a man named Price, author of the Price equation about gene changes in frequency over time, has done quite a bit to bolster confidence in Fisher's results.

The field of mathematical evolution is quite popular with scientists today. Most of the theorems in this field assume the existence of a replicating population (entities that replicate themselves from generation to generation with minor random changes) that compete for resources, and whose success in competition correlates positively with the number of their offspring in the next generation. The inference is that evolution takes place in such a population. Because evolutionary theory is based on proven mathematical laws, to verify the parts that have been described mathematically, all we need is to show that the conditions of the mathematical theorems are satisfied by nature. If we find that these conditions are met, we must conclude that evolution is taking place in this population. Research shows that the conditions are indeed met. The replicating elements are genes or whole creatures, in which replication with minor changes and competition for resources indeed occurs, and therefore they evolve. This conclusion is inescapable.

The existence of evolution in nature has been confirmed in many ways, whether by observations of fossils and animals, by experimenting

---

[99] Gene frequency.
[100] Fisher's theorem.

with the creation of new species,[101] by following the development of organisms in nature,[102] by examining the genomes of various organisms,[103] or by resistance to antibiotics developed by bacteria and resistance to insecticides developed by insects. Our ancestors already relied on evolution, without being aware of it, when they domesticated and improved animals and plants, and when they attempted to improve the race by making informed choices of spouses.

The most impressive confirmation, in my opinion, is the one derived from the protein structure. Evolutionary theory cannot predict what the next stage in the development of a living being will be, but our knowledge of genetics enables us to guess what the previous stages in the evolution of existing beings were. The structure of the proteins allows us to measure the evolutionary distance between two different versions (in two different living beings) of a given protein, and infer from that distance the place of their common ancestor down the evolutionary tree. This ability makes it possible for us to discover the evolutionary tree structure that links different animals by comparing the different versions of that protein. This satisfies those who believe that evolution takes place, but someone who rejects evolution may argue that if we tested another protein, we would get a different tree. To show that this was not the case, scientists examined the evolutionary tree structure of 11 species of animals obtained from the analysis of five different proteins. The trees obtained in the experiment were very similar. The experiment confirmed at the same time both the existence of evolution and the way the evolutionary tree was constructed according to the proteins. The odds of such a high correlation between the various trees would have been close to nil if evolutionary theory had not been correct.[104]

As noted, these are confirmations of the fact that evolution exists in the animal world, which is unrelated to the fact that the mathematical theorems are correct without resorting to the world of living creatures.

---

[101] Example of experimentation in creating new species.

[102] Example of the creation of a new type of lizard.

[103] Example of a comparison between the human and chimpanzee genomes.

[104] Description of the experiment in *Nature*. The full text of this article can be read free of charge here.

But because these mathematical laws do not depend on the existence of living replicators, evolution has also been used to develop software and robots. This is done, among others, by means of evolutionary algorithms.[105] There have been considerable achievements in this field, as mathematical laws not only explain, but also create in practice, physical reality. Among these we can find software that plays chess,[106] evolutionary robots,[107] an automatic scientist,[108] and more.

Naturally, these are not proofs of mathematical theorems, as these theorems have been independently demonstrated mathematically. They are technological applications of the theorems that artificially create the conditions in which the theorems show that evolution occurs, in order to obtain evolution.

Another application that exploits and confirms the theory is the development of drugs by an evolutionary process of selecting variants of chemicals that increasingly excel in dealing with diseases or toxins.[109]

6.2 Intelligent Evolution?

Some proponents of intelligent design cannot ignore the fact that there is such a thing as evolution, and to protect their faith against obvious facts, they claim that evolution is directed by some intelligence.

Evolutionists know that this is wrong. There is no shortage of articles trying to establish the argument that evolution operates randomly, without any guidance. Some of these articles point to extinct species and ask what wisdom can stand behind developing a creature destined to become extinct. Other articles point to strange and superfluous properties of humans and other animals as proof of poor planning.[110] They raise

---

[105] Evolutionary algorithms.
[106] Software that plays chess.
[107] Robots developed in an evolutionary process.
[108] Automatic scientist and more.
[109] Development of drugs in an evolutionary process. Free copy of the same article.
[110] Poor planning.

questions like: What do ostriches need wings for? Why do the embryos of whales have legs that subsequently atrophy?[111]

Although these arguments are appropriate for opposing standard creationism, according to which living beings were created "as is," and are not the result of evolution, they fail to refute evolution directed by intelligence. Creationists can claim that extinct species or superfluous traits are like scaffolds. The houses we live in are also the result of intelligent planning, but nevertheless, during construction, we find scaffolding and supports all around the building, which we don't find in the finished house. Creationists can argue[112] that dinosaurs were needed as a stage in the evolutionary development of later life forms. They can claim that the best path to developing the ostrich was through birds, and later, the intelligence may remove the wings. And so on.

Note that it is very difficult (in fact, it is impossible) to refute these arguments because they rely on the true course of evolution, adding to it only the claim that everything happened deliberately. For this reason, I prefer to base my opposition to intelligent evolutionary theory on another argument.

Evolution works through mutations. The accepted view in science is that mutations are random, but proponents of intelligent evolution claim that these mutations are intentionally created. I suggest that you think first of all about the degree of intelligence necessary to direct evolution. This intelligence must choose the mutations that would advance its purpose (there's no planning without a purpose). To plan mutations, it is necessary to be able not only to predict the effect of any given mutation, but also to detect the mutation that would make possible progress toward a certain goal. This is a level of intelligence and knowledge that humanity, despite years of research, is still far from achieving. In other words, this is really (but really!) supreme intelligence.

---

[111] Why do the embryos of whales have legs that subsequently atrophy.

[112] I have not come across a creationist who made this argument, but intellectual honesty requires me to point out a failure in the argument even if that argument is intended to support the position I hold.

Now, think about the nonsense that evolution perpetrates every day. For example, it repeatedly produces people with Down syndrome.[113] Why does it do that? After all, this is clearly not a scaffold for achieving a more advanced version of humans, but a pure malfunction. OK, you may say that even a superior intelligence can be wrong here and there. But to repeat the same mistake over and over again? This is already the definition of supreme stupidity![114]

6.3 Evolution, Intelligence, Planning, and Semantics

There is a certain similarity between the products of intelligence and those of evolution. Does this similarity justify the description of evolution as "the intelligence of nature"?

The story "6.2 Intelligent Evolution?" rejects the claim that evolution is guided by an intelligence. The present story discusses the claim that evolution itself is some kind of intelligence.

These are two distinct claims because there can be intelligent entities that are not guided by intelligence. One such entity is our brain. Its thinking process often creates nonsensical ideas, not unlike evolution, which also produces nonsensical mutations. Therefore, we may reach the conclusion that neither the brain nor the mutation process are guided by an external intelligence, and that the brain then discards the nonsensical ideas just like natural selection discards nonsensical mutations. Given such similarity between evolution and the processing performed by the brain, is it justified to claim that the brain is intelligent while evolution is not? I discuss this question below.

We spoke about the resemblance between evolution and intelligence in their outcomes (or products), but this resemblance has another dimension, which is the source of these outcomes. In many ways, evolution applies the scientific method: it conducts experiments by creating a

---

[113]This is just an example of a whole series of malfunctions caused by trisomy mutations.
[114] Supreme stupidity.

mutation, and upholds or rejects the assumption of the experiment (that the mutation is useful) through natural selection. This similarity between evolution and intelligence is familiar to us. But is there anything that distinguishes evolution from intelligence?

How do we identify intelligence in organisms? The main feature we expect to see in the behavior of an organism to which we ascribe some (even if only a little) intelligence, is goal orientation. Behavior of this type seeks to achieve overarching objectives by forming and achieving intermediate goals. It does not respond to events identically and automatically, as would event-driven behavior, but considers the response to them in a wider context of achieving a specified goal.

It follows that to be able to identify intelligence in any organism, we must identify its goals, or at least we must be persuaded that it has goals. Does evolution have goals? I don't see how any could be identified, but this in itself doesn't mean that it has no goals.[115] Nevertheless, I argue that evolution appears to have no goal of any type.[116] I make this argument not because we cannot recognize such a goal, but because in my opinion to talk about the goal of evolution (or of God, or of anything that is not a machine) has no meaning. As far as we know, the sense of purpose we feel is merely an expression of a certain activity in our brains. In other words, it is a neurological phenomenon, and at a certain level of generalization, a physical one. This is also the case with the artificial intelligence systems we build; they are physical systems, in which the goal is also defined physically. Attributing a sense of purpose to an entity that is not physical

---

[115] The excuse "the ways of God are inscrutable" has been worn out by overuse, but it, too, can be interpreted to mean "there are goals, but I don't know what they are," or in other words, "you cannot contradict what I just said because I didn't say anything." For some reason, those who raise such an argument understand that it is not possible to contradict them because they haven't said anything, but cannot understand that for exactly the same reason I cannot accept their argument.

[116] A goal is a kind of *quale* (a sense of something). I don't attribute to the falling ball the goal of "getting down," although I know that this is what it will do.

is as meaningful as the statement that such an entity can feel fear, help-lessness, or a headache.[117]

Is it correct, then, to define evolution as a kind of intelligence? This is merely a semantic question that says nothing about the intelligence of living beings or of evolution. The whole function of the question is to determine the meaning of the word "intelligence." I prefer not to describe evolution as an intelligence because the difference between the intelligence we know and evolution, which is aimless, is too great, and using the same description may be confusing.

---

[117] Note that the feeling of purpose, just like fear, is a quale. Evolution is not a physical entity, and the only qualia we have ever encountered are representations of patterns of neural activity in brains. We don't even know whether we all experience them in the same way (see "1.3 *Qualia: The Limits of Understanding*"). This means that when we attribute the "pure qualia" of purpose to evolution (qualia that are not representations of neural activity that we don't even know can exist), each of us may be talking about a different thing.

Chapter 7

Other Minds

We address the question "What can we expect from man-made intelligence?" To what extent is such intelligence similar to ours, and what are the advantages and disadvantages of complete vs. only partial similarity?

7.1 Bodiless Horseman: Does the Mind Need a Body?

We often hear claims by intelligence researchers and the interested public that there can be no mind without a body. I don't share this position.

Philosophers of yore saw the mind as a completely independent thing that exists separately from our physical body. Scientific research has shown that this is not so, and that all components of our soul, including our intelligence, are the products of our body, especially of our brain.[118] This conclusion, which began to take shape even before conclusive research evidence was available, has led many scientists to try to develop artificial intelligence using computers.[119]

---

[118] These conclusions are discussed in several stories in "Chapter 1. Soul, Mind, Will, and Self (Are) Matter."

[119] The term "computer" refers here to all types of automated systems used to develop artificial intelligence, including neural networks.

Following the limited success of existing computerized systems in demonstrating real intelligence, some researchers have begun to argue that the conceptual integration required is broader than previously thought, and that the infrastructure necessary for intelligence includes the entire body, with all its motor and sensory abilities, not only the brain. For example, an article published by NCBI[120] claims that "the growing discipline of embodied cognition suggests that to understand the world, we must experience the world." In another article, published in *Science News*,[121] we read that "many people today do not realize, though, that there's a modern version of Descartes's mistaken dichotomy. Just as he erroneously believed the mind was distinct from the brain, some scientists have mistakenly conceived of the brain as distinct from the body."

The motivation behind this position is somewhat similar to that behind the story "3.6 Philosophical Movement: Do Plants Feel Pain?" in which I explained why I don't believe that plants have consciousness. But this position ignores the fact that the brains we are familiar with have developed in an evolutionary process.

I believe that this position is mistaken, and that it is possible to construct a thinking, disembodied brain. I'm not talking about a brain that has no way of interacting with the environment. (Such a brain may also be possible, but it would not be able to learn or teach anything, and existing in isolation, it may suffer a great deal, as experiments in sensory deprivation suggest.[122]) What I have in mind is a brain that can receive information and share conclusions with us through input and output.

One indication that such a brain is possible comes from people who have been completely paralyzed, and in practice have lost all contact with most of their body, although their consciousness has been preserved.[123] The film *Matrix* also appears to illustrate a hypothetical situation in which the body is completely neutralized, and all sensory input is received from a computerized system that simply makes it up.

---

[120] Does intelligence require a body?

[121] To have a sound mind, a brain needs a body.

[122] A human brain that has no interaction with the world is liable to suffer a great deal, as experiments in sensory deprivation suggest.

[123] Read about quadriplegia and locked-in syndrome.

I have no doubt that a body was required for the brain to develop in the course of evolution, as I mentioned in "3.6 Philosophical Movement: Do Plants Feel Pain?"; indeed, a body that can move. But contrary to the view that every brain necessarily needs a body, in my opinion the body is needed only as part of the evolutionary package, so to speak, and not in principle. It is also possible to argue that to be able to understand anything, the brain must be equipped with "atoms of understanding" of the type I discussed in the story about *qualia*. But although the qualia are created, among others, as a result of the interaction of the body with the world, they may be built into an artificial brain, without such interaction. Such a construct may be created, for example, by accurately copying the parts of an adult human brain to a neural network. It follows, therefore, that I'm a believer in the human brain project, which attempts to create, as its final product, a body-less electronic brain that simulates the human brain well enough to make it possible to use it to study neurological and mental disorders.

7.2 Will Frankenstein Rise Against Its Creator?

Does progress in artificial intelligence necessarily lead to a rebellion of machines against humans?

At the end of the story "1.4 I, Me, and Myself," I hinted that the danger of machines taking over the human race, which many fear, is not as great as filmmakers would have us believe. My confidence is based on the fact that the "self," as well as the anger, jealousy, hatred, desire for vengeance, etc., are all products of the evolutionary process, in which the degree of success in the competition for resources is what selects the survivors. In other words, all the qualities that cause us to fight each other and to lord over other species are the result of the cruelty of natural selection rather than of intelligence. Therefore, the machines we are likely to produce do not have to be as selfish as we are, and if we act wisely, they will not try at all to control us, not because they won't be able to, but because they will not be motivated to do so.[124]

Yet, a few reservations are in order.

---

[124] AI takeover.

7.2.1 *First reservation*

Even if intelligence in itself is not expected to cause a confrontation between us and the machines, other traits that we perceive as positive may lead to such a confrontation. Prominent among them is empathy, which is based on our ability to feel what others feel, including anger and other negative emotions. The contradiction between empathic ability and the ability to live with others in peace does not stem from the fact that empathy developed in the course of evolution, but from the fact that empathy is the ability to feel what others feel. Therefore, if you want the machine to avoid acting in ways that will make us angry, it must understand the concept of anger in us, that is, it must be able to get angry. The same is true regarding other negative feelings. Thus, even if there is no contradiction between the intelligence of a machine and its lack of motivation to gain control of humankind, there is liable to be[125] a contradiction between the empathic ability of a machine and its ability to live with us in peace.

7.2.2 *Second reservation*

If the machines develop without evolution, it appears to be rather simple to avoid their becoming egocentric. Although it is not necessary for intelligent machines to develop through an evolutionary process, it could accelerate the development of artificial intelligence, making the machines partners in planning and creating next generations of even more intelligent machines. Therefore, a contradiction may emerge between the ability of machines to live with us in peace and the byproducts of the evolutionary process by which we instilled their intelligence in them.

Nevertheless, not every guided process of evolution will necessarily generate violent tendencies. This depends, among others, on the resources

---

[125] Notice the term "liable to be..." The contradiction is not entirely certain. It is possible, for example, that we can circumvent the problem if we teach machines to understand our brain as a machine, so that they can identify the sets of activities within it that are identified with this or that *qualium*, and even predict their development. The applicability of this idea depends, among others, on the degree of similarity between the activity of our brains. In such a case, it would be possible to develop in machines "love" or "aversion" of certain activities in our brain without their experiencing the *qualia* that we experience.

for which the products of evolution are competing. For example, if the resource is human preferment (that is, humans will determine according to their preferences which systems are entitled to have offspring), it is unlikely that violent tendencies will emerge.

Chapter 8

How Come the Universe, Nature, and Life Exist?: All Beginnings are Difficult

In the story that considers whether we will ever know everything, I explained that there will always be questions that we cannot fully answer. This is not something to discourage us, however, and we continue to look for the reasons behind things, although it is clear to us that when we find such reasons, the question immediately arises as to the reasons for these reasons. In this chapter, I present some of the possible answers to questions about the reasons why things exist.

Questions of the type "How did a certain natural phenomenon begin?" were always difficult to answer for two reasons:

1. They always concern the distant past, and there is no evidence to confirm this or that conjecture.
2. Even when it is possible to point to a possible reason for a beginning, the question always arises "What is the reason for the reason?" (See "3.3 Will We Ever Know It All?")

Religious preachers use such questions because they know that these are places where the "God of the gaps" can still find refuge, as science has not yet answered them indisputably. But we know that the God of the gaps is a rather foolish idea, mainly for two reasons:

1. Even if science has not yet answered the question, it does not follow that the answer is God, and certainly it does not follow that the answer is the God of any particular religion.
2. Assuming the existence of an entity capable of creating universes requires itself an explanation, or at least some confirmation of the correctness of the assumption.

Religious preachers should be shown that in many of the questions they raise, the situation is far from similar to the one they imply by their question: it is not that science doesn't know how things could have happened, but that out of the vast array of possibilities, including those known to science and those not yet known, there is currently no way to determine which one describes what actually took place.

Therefore, many of these are not questions of principle, but historical ones, that is: questions of what actually happened. It is not clear whether we will ever have the data needed to decide between the alternatives. But even if the data are never found, we can already discard the God of the gaps, simply because the explanation has no need for him.

I briefly review below some of the questions of this type, and point to solutions that science offers to them without resorting to God.

8.1 Why Did the Big Bang Occur?

Several conjectures have already been suggested as a possible description of the process that caused the Big Bang. It would be impossible to cover all of them here. The text of this story should not be regarded as a full account on the possible causes leading to the Big Bang. This story is included in the book merely to show that the scientific community does not rely on divine intervention to kick-start the universe. Therefore, I describe only briefly some of the main ideas appearing in these conjectures. For a deeper understanding, I encourage you to read the books mentioned in this description.

The proposed causes for the Big Bang can generally be grouped based on the following ideas:

1. Time, despite its finality, need not necessarily have a beginning. It is rather like a ball, an idea presented by Hawking in his book *A Brief History of Time*.
2. Our universe is part of a "multiverse," in which new universes are constantly created, cyclically or simultaneously.
3. The universe is merely an expression of mathematics, as proposed by Max Tegmark, under the name of "mathematical universe hypothesis."[126] According to this assumption, the universe is not only successfully described by mathematics, but is itself a mathematical structure. The very existence of something is, therefore, based on its mathematical coherence.
4. The universe was created from a deeper nothing, where even space does not exist, and it may return to nothing. The idea of "a universe from nothing" is described by Lawrence M. Krauss.
5. A cyclical universe with timeless "segments," proposed by Roger Penrose in his book, *Cycles of Time*. Penrose's proposal relates the existence of time to that of mass. It does so based on two mathematical expressions of energy. In the first expression, energy is proportional to the frequency, and in the second, it is proportional to the mass — suggesting that the existence of a frequency (necessary for measuring time) is equivalent to the existence of mass.

   This fits well with the theory of relativity, according to which massless particles must travel at the speed of light, and therefore do not "feel" the presence of time. Penrose assumes that given enough time, everything that exists in the universe will be transformed into massless particles, time will be lost, and consequently the dimension of distance will also be lost (from the point of view of the particles), recreating the conditions in which the universe was created.

---

[126] Mathematical universe hypothesis.

6. One of the intriguing features of this idea is that it copes successfully with the fact that for entropy to grow over the life of the universe, it must start from a lower value than the one we currently encounter. What creates the potential for the growth of entropy in every cycle of the universe is the formation of mass following the Big Bang, at the beginning of that cycle. This solution also obviates the need for the inflation mechanism that was proposed to explain the size and uniformity of the universe, because according to Penrose, the universe has inherited its current size from the earlier universes, each of which began with the size of its predecessor.[127]

8.2 Abiogenesis: Kick-starting Evolution

Several possible answers were also offered for the abiogenetic question of how life first emerged from inanimate matter, which fit into the overall picture of evolution. From time to time, in the course of an argument between adepts of science and creationists about whether or not evolution exists, the creationist springs the "crushing" question about the formation of life. Many proponents of evolution tend to answer that this question is not part of the theory of evolution. Doing so, one effectively concedes one's position in the debate because one's opponent maintains his conclusion that "evolution or not evolution, no matter what you call it, abiogenesis is something essential that requires a Creator for the world."

It is a shame to lose the argument on this point (after all, the objective of the creationist is to show that God is necessary, without caring much for what purpose exactly), because the claim that the abiogenetic question is not a part of the theory of evolution is not accurate. We have ample evidence that from the moment RNA-based life has been created, evolution has occurred, and that all known life forms are its consequence. Moreover, the only rational explanations we have for the creation of life from inorganic matter, and its development to its present state, are based

---

[127] The idea is described succinctly in the video, "Why did our universe begin?"

on evolution. Although we have no precise information about how this process actually happened, and perhaps we never will, all the speculative explanations that exist today are based on evolution that begins with the spontaneous creation of some replicator (which is not based on RNA but on much simpler structures), and on the replication of this replicator, with increasing sophistication, until the formation of RNA.[128]

Both creationists and proponents of evolution sometimes take refuge in answering the question of the formation of life in the theory of panspermia.[129] This "explanation," however, does not relate to the question of *how* life was created, but only to the question of *where* it was created.

A different but related subject concerns the question "Is life a result of chemistry and physics?" It is possible to obtain a hint to the answer from experiments conducted by Craig Venter, in which he created a bacterium that did not exist in nature, in a process in which most steps were completely synthetic. From the cell of a living microbe, Venter removed all the hereditary material (DNA), and inserted a different hereditary material into it, which he created in the laboratory, using conventional chemical processes.[130]

8.3 First Love: How Did Sexual Reproduction Begin?

The argument of irreducible complexity (IC) is one the creationists' favorites, and sexual reproduction is one of the examples they often use.

---

[128] For an article published in *Science* that talks about the creation of the building blocks of life, see <u>Researchers may have solved origin-of-life conundrum</u>. In a more recent research, scientists developed a software tool that constructed many possible routes from the prebiotic world to abiogenesis. The software tool (<u>Alchemy</u>) has been made publicly available. Its description can be found <u>here</u> and <u>here</u>.

[129] The hypothesis that life exists throughout the universe, and has been transferred to Earth by space dust, asteroids, or other vehicles.

[130] See <u>Craig Venter's lecture</u> for more details.

The main idea of contemporary evolutionary theory is that the variety of life is the product of a process whose main elements are mutations (small changes in the genome) and natural selection. Mutations are the "creative" force of evolution that brings new creatures into play. Natural selection is the "audit system" that removes from the board creatures that do not fit their environment.

Many important traits of different organisms are the result of the combined activity of several genes. The IC argument is that some of these traits require a series of mutations, none of which in itself provides the organism with an advantage. Therefore, the chances of all of them meeting in the same genome and providing the organism with a trait that they can provide only if they all exist together is next to null.

Note that the IC argument is not a serious one, because it is an argument from ignorance ("I don't know how to reduce the complexity of this trait, and therefore it's irreducible"). Claims relying on this argument should be addressed, nevertheless, because creationists keep raising them and because exploring them increases our understanding of nature.

IC claims have been raised against the evolution of various traits in the animal world. A well-known example is the eye. Scientific research has discovered that not only is the evolution of the eye reducible, but that it has happened in nature several times independently. There is even evidence of intermediate stages, starting from no sensitivity to light through photosensitive cells, and eventually to fully developed eyes.

The possibility of claiming IC appears to increase exponentially when the complex feature requires reciprocity, so that it grants its owner an advantage only if other beings have a corresponding trait. Examples of such features (which, to the best of my knowledge, were overlooked by creationists) and of my suggestion of how to deal with them can be found in stories on the development of language and on empathy and altruism.

An example of yet another such feature is sexual reproduction. As in the case of other mutual traits, explanations have been suggested for how this ability survives in evolution, but it is more difficult to explain how it came about in the first place. Scientific literature contains speculations about how a sex-like process such as transgenesis may develop in bacteria because of group selection, and about the advantages transgenesis can provide by speeding the rate of propagation of useful mutations, but I'm not aware of any attempt to explain the evolution of sexual reproduction in multicellular organisms. Creationists used this gap in understanding to insert the God of the gaps through it. Quite a few articles have presented sexual reproduction as a prime example of IC, which refutes the claim that evolution is responsible for the development of all traits.

In the current story, I intend to close this gap and show how sexual reproduction can be created in an evolutionary process without the involvement of an imaginary friend. Note first that, contrary to multicellular organisms, in single-cell organisms group selection is often a satisfactory explanation for the development of traits that require reciprocity for providing an advantage. This is because a mutation in a single organism is sufficient for a large group of organisms to be present shortly thereafter in the same place, carrying the same mutation (and therefore able to benefit from the advantage that the mutation provides). Therefore, explanations of the existence of gene transfer between bacteria (transgenesis) based on group selection are quite reasonable.

What happened when multicellular beings were first formed? To answer this question, we must ask how the first multicellular organisms procreated. Apparently, cells that had been shed or secreted by them evolved into new multicellular organisms, just like the first multicellular organism of their type evolved from a single cell. If, therefore, the multicellular organism evolved from a single-cell organism that is capable of exchanging genes with other single-cell organisms of its kind, the cells that have been excreted from the multicellular organism and developed into a new multicellular organism must also have this trait.

With the "invention" of the process of tissue differentiation, a tissue whose function is to secrete reproductive cells has been preserved. Subsequently, organisms evolved whose reproductive cells specialized in introducing their genes into other cells (males), and organisms whose reproductive cells specialized in receiving the genes from others (females). Males who maintained maximum closeness to the females during the secretion of the reproductive cells had an advantage. "Identifying" the advantage of fertilizing the female cells as close to the source as possible eventually led to internal fertilization. The fact that even present-day multicellular organisms reproduce by single-cell reproductive cells supports the story. It is possible to claim that, at the most basic level, we are colonies of single-cell organisms (reproductive cells) that developed additional organs. Note that the process proposed here, of the development of the multicellular offspring from a reproductive cell in imitation of the creation of the first multicellular being, appears to be the seed (literally and figuratively) of the phenomenon called recapitulation theory.[131]

8.4 Divine Symmetry

Symmetry as a "force" of creation and design in nature.

In "3.4 Meta-beauty" I explain that the simplicity of symmetry plays a significant role in our perception of beauty. In this story, I'd like to point out some of the more miraculous and less familiar roles that symmetry plays in the world.

A few years ago, in an online debate on evolution, an anti-evolution religious antagonist challenged me with the following question: "How can you explain the fact that evolution randomly created symmetry between the two sides of the body of all animals?" My answer was simple: Symmetry need not be the result of evolution because it is the simplest and most natural state. It is precisely lack of symmetry that requires an evolutionary explanation.

I believe that the first multicellular creatures on Earth had spherical symmetry, so that they were much more symmetrical than the creatures whose symmetry my opponent wished me to explain. This assumption stems from the fact that the development of cells is guided by their chemical environment, and their chemical secretions diffuse symmetrically in all directions. For the same reason, all the symmetries that have not been

---

[131] Recapitulation theory. The hypothesis that the development of the embryo of an animal represents successive stages in the evolution of the animal's remote ancestors. Although this theory has been largely disproved, some of it is still considered correct. See the Modern Status section of this entry.

broken are preserved in living organisms.[132] Even the complex animals we are familiar with today begin their development as a symmetrical block of cells.[133] The phenomenon that we identify as the reflection symmetry of animals is merely an expression of the partial breaking of the rotational symmetry as a result of environmental conditions that have affected evolution.

In the animal kingdom, the first development, apparently, was the creation of a difference between "front" and "back," with the emergence of the mouth and the digestive system. Naturally, emergence of the mouth also encouraged the development of various sensory organs in its environment. The next difference was probably the one between "up" and "down." The environmental pressure to develop this difference was triggered by the force of gravity, which pulls downward only.[134] The environment has never exerted any pressure on animals to encourage the development of differences between "right" and "left," which is why we find two-way symmetry in animals.

Up to this point, we discussed symmetry only in the highly familiar sense of the word, the types that are likely to exist in different directions in space. Science, however, also deals with other symmetries. See the definition of the term in Wikipedia, according to which symmetry exists when there are characteristics of nature that a certain type of change in the environment does not affect. The type of change in the environment relevant to this or that symmetry may be a change in the location in which the

---

[132] For the sake of simplicity, let us imagine a two-dimensional creature whose cells populate part of the cells of a matrix. Consider an initial sequence of different cells (say, A, B, C, D, E, F, G, H) going from front to back (the direction in which the symmetry was broken), and a set of rules (imprinted on their DNA) that dictates whether or not to grow a neighbor in the vacant places on their right and left. Consider also a collection of additional rules that tell the cell how to differentiate (into what type of cell to turn — A, B, C, etc.) according to the identity of its neighbors at the moment of its creation. These rules are also stored in the DNA. Now run the rules on the system each time anew, that is, on the initial row, then on the collection of cells created adjacent to it, and again on the collection of cells created next to them, and so on. The shape you obtain will necessarily have reflection (two-way) symmetry.

[133] Morula.

[134] This development also has an echo in the developmental process of the fetus. For more on the subject, read the story "11.2 Symmetry-Breaking Evolution and a Wild Hypothesis" in Chapter 11.

testing of the characteristic is being performed, in the direction in which the evaluator looks, in the time when the evaluation is performed, and more. One of the most significant contributions to understanding the way in which different symmetries influence nature is the work of the German mathematician and physicist Emmy Noether, whose acceptance by the scientific community, despite her being a woman and a Jew, is a fascinating story in itself. The contribution to which I refer here is the formulation and proof of Noether's theorem, which makes the connection between the symmetry of a physical system to the conservation laws that apply to it. This theorem reveals, for example, the following facts:

The law of conservation of linear momentum derives from the symmetry of spatial displacement;

The law of conservation of energy derives from the symmetry of temporal displacement;

The law of conservation of angular momentum derives from the symmetry between the various directions;

Etc.

In various disputes with religious opponents, the argument is often raised that if there are laws in nature, they must have a legislator, and that legislator is God. I discuss the illogical part of this argument in "10.2 I Am Who I Am: Paradoxes of Self-reference," but here I'd like to address this claim from a different direction, and relate it to another claim (this time, an atheist one) that arises in theological discussions: that if God is perfect, He has no reason to create a world because nothing is missing.

Note that symmetry can be described as an expression of indifference:

"I don't care where you conduct the measurement"

"I don't care when you conduct the measurement"

"I don't care in which direction you perform the measurement"

And so on...

In other words, for lack of an alternative, many natural laws are the product of indifference, which can arise only in one of two situations:

Either there is no God, or there is a God[135] who doesn't care, and laws are created without any involvement on his part. This story is called "divine symmetry" because it shows that symmetry plays part of the role ascribed to God.

8.5 Die and Let Live: About the Birth and Life of Death

Why we age and die has troubled humans since time immemorial. Does evolution help answer this question?

Many scientists provide only a partial answer to this question, which I think misses the point. Their answer is that the body "wears out" over time, and from the point of view of evolution, "there is no point in investing"[136] in repairing malfunctions after the organism ceases to be fertile. Because at this point the organism no longer contributes to the reproduction of its genes, the defect repair mechanisms stop working. Moreover, defects that are by nature cumulative (such as defects in the defect repair mechanism) are not corrected even at a young age, if the rate of their accumulation results in a serious disorder only after the period of fertility has ended.

These are, of course, correct considerations, but they miss the main point because they consider the limitation of the age of fertility as something naturally foreordained that doesn't require explanation. It is quite clear that the reproductive mechanisms cease to function for similar reasons of accumulation of damage and a failure of the repair mechanisms. But why doesn't evolution "fight" these phenomena, while gradually

---

[135] Whatever it may be… Indeed, we tend to reject the existence of an ineffective God based on Occam's razor.

[136] Evolution obviously doesn't "invest" in anything. Throughout the story, I use a teleological language that ascribes purpose and intention to things, but I do so only because it allows short and clear figures of speech. In practice, things happen without a guiding intention, and the rationale presented as their motive merely represents the fact that, in retrospect, of all the things that have evolved randomly, those that were chosen by evolution have been given priority because of this rationale. For the teleological formulation not to mislead, I occasionally use quotation marks to emphasize the fact that this is merely a turn of phrase.

validating more effective damage repair in other systems as well? To answer this question based on evolution, two questions must be asked: (a) whether there is a reason for evolution to "push" for immortality in the first place, and (b) whether populations of immortal or extremely long-lived individuals are likely to survive over time. The following two sections address these questions.

8.5.1 *Does evolution have a reason to make a push for immortality?*

I argue that the answer is negative. Evolution uses what programmers call a quick and dirty method. It doesn't plan its actions, and if a feature is beneficial for survival of the genes, even at the price of the mortality of the organism, it is given preference. Most of the multicellular dynasties started out as mortal through the accumulation of beneficial mutations that do not preserve immortality. When we understand this, we realize that in dynasties that started out as mortal, immortality could have evolved only if it was truly beneficial to the survival of the genes.

This is perhaps the place to clarify a point that is often overlooked: when we talk about the "selfish gene" that "is concerned" with its own survival, we do not talk about a concrete, material instance of the gene. What these expressions aim at is the structure of the gene. Evolution selects the structures of the genes rather than their concrete instantiations. Therefore, the question of immortality is almost illegitimate, because it appears to mobilize the evolutionary pressure to preserve the structure as a justification for the preservation of concrete instances. In reality, the successful structures are truly immortal, and what dies are the concrete instances.

To examine the "attractiveness" of immortality from the point of view of evolution, two questions must be answered: "What is the probability for concrete instances of the gene to randomly develop immortality?"[137] and "To what extent is the immortality of concrete instances of a gene

---

[137] In other words, of the organism of which the concrete instances are a part.

beneficial for the survival of the gene structure, which is, as noted, the object of natural selection?"

The answer to the first question is that this probability is extremely low, among others because the differentiation of the tissues[138] (their specialization in specific activities) is based on a change in the expression of their genes, and there is no factor beyond natural selection that "stands guard" to watch that any such change does not harm immortality. As tissues differentiate into more and more types, the "attractiveness" of repairing malfunctions in each tissue diminishes, to the point of vanishing altogether. This is because the life of the organism is not extended if only some of its tissues become immortal, and as the number of tissue types increases, the chance of a combination of mutations that would repair all of them approaches zero.

This low probability highlights the importance of the second question: To what degree does the immortality of concrete instances of a gene benefit the survival of the gene structure? The benefit is not great. Imagine that over a given period of time an organism gives life to N surviving offspring. If, during the same period, it gave life to N + 1 surviving offspring, it could have comfortably died after this period and at the same time achieve all the benefits of immortality from the point of view of gene structure.[139] In other words, the advantage that immortality grants living creatures in the survival of their genes is identical with the advantage of bringing one other surviving offspring into the world, and no more than that. Therefore, the path of giving life to additional surviving offspring, which is not limited to a single additional one, is a much more "attractive" path for evolution.

---

[138] Which is one of the important differences between multicellular creatures and a random collection of cells.

[139] For example, if the given period is one year and N = 2, assume that organism A is immortal and produces two offspring a year. Such an organism triples its quantity each year because each single organism is replaced by three (itself and its two offspring). Now assume that organism B lives for only one year but produces three offspring during that year. This organism also triples its quantity each year, which makes it, from the perspective of the genes, equivalent to A.

8.5.2 *Why is immortality mortal?*

Even in the rare cases in which a certain population is close to achieving the immortality of its individuals, this population is likely to eventually become extinct. Bacteria, for example, do not die of old age. They die of hunger or are killed by antibiotics, but not of old age.[140] Then who dies of old age? It seems to me that death by old age was invented by larger animals. There is good reason to believe that, beyond the absence of an evolutionary reason to prevent it, death also has to do with size — not in the sense that "size kills," but because it allows (indeed, requires) differences in size between old and young, and because of the struggle for resources. Size does matter, after all. Larger is also stronger, and in situations of scarce resources, the adults, were it not for aging and dying, would almost always prevail over their offspring. Such a reality would make age a more significant factor in survival than genetic adaptation to the environment. Because the genetic diversity created in the reproductive process is the only thing that can save the species in the case of a drastic change in the environment, a significant reduction in the effect of genes on the survival of individuals significantly reduces the chances of survival of the species when such a change occurs.[141,142]

---

[140] Unless we define the division of bacteria (binary fission) as the death of the divided bacterium. Such a definition may be justified, because it is difficult to determine which of the two bacteria created is the "original" bacterium, and which is the offspring. At the same time, it is difficult to define the event as death by old age because it is not the result of wearing out.

[141] Note the similarity between this situation in the animal world and that of companies that excessively prefer veteran employees over talented ones. Such companies become corrupt, and unless they are supported by the government, they also die.

[142] Genetic adaptation to the environment is usually a continuous and gradual process, and often changes in the environment are of the same type. In any population, there are several individuals more resistant to the emerging conditions, and they enjoy an advantage that enables them to produce offspring, some of which will be even more resilient. This is why physicians advise taking the full dose of antibiotics even if we start feeling better before completing the course of treatment. The idea is to kill even the few bacteria that were able to survive the partial dose, to prevent the development of their even more resistant strain. Excessive preference given to the age of the organism over its genetic suitability impairs its ability to gradually develop resistance to new conditions.

The explanation that I suggest is therefore that species in which there is a size difference between parents and offspring need death to survive. Aging and death are necessary to prevent age from being the primary factor affecting the chances of survival, and to ensure that genetic compatibility with the environment holds center stage. A species that overcomes death, or moves too close to immortality, would eventually become extinct.[143] In other words, limitations on the age of fertility are not the reason why evolution chose not to "fight" the aging of other systems. The aging and death of the organism do not harm, indeed, they contribute to the chances of survival of the species, and limiting the age of fertility is simply part of aging.

All of the above suggests that in the vast majority of multicellular species, death is not a law of nature in itself, but the result of a combination of an absence of evolutionary pressure in the direction of immortality and a certain type of group selection. One question still arises: "Why do parents, who are generally so concerned about their offspring, fight them for resources?" My answer is that parents in nature are concerned about their offspring, but not to the point of sacrificing life, because if an offspring needs the parent's care to survive (and there is a correlation between need and concern, because once the offspring becomes independent, the parent stops showing concern), the death of the parent will in any case result in the death of the offspring as well. Therefore, self-sacrifice makes no sense. A clear example of this phenomenon is the behavior of female tigers when a male attacks and kills their cubs. Although the females are generally extremely caring of their cubs, they do not sacrifice themselves fighting for their lives. They even allow the attacking male to have his way, and go into heat after the death of the cub.

---

[143]There are exceptions. Hydrozoa, for example, almost don't age, and there are even species that probably do not age at all. Yeasts, however, are single-celled organisms that do age and die, but in their case, there are also differences in size between parent and offspring. Hydrozoa live in seas and lakes, which are relatively stable environments. Even if there is a change in the environment, a slight vertical movement usually allows a return to an environment in which the conditions are similar to those preceding the change. These exceptions seem to support the correctness of the hypothesis.

Note further that trees can live much more than other forms of life. This makes sense, given that their ability to compete with their offspring is lower because they are not mobile. For the same reason, trees can "afford" to have even much larger size differences between adults and offspring than animals can.

As human beings, we try to fight death, but the more successful we are at it, the more we will have to contend with the need for the individual to die to enable the survival of the species. This is not too bad; we have a good chance of succeeding in coping with the problem, because unlike other animals, we can use our intelligence to adapt to new conditions, and need not rely on the chance of getting the right mutation.

8.6 Do We Live in the Matrix?

The most conspiratorial of conspiracy theories.[144]

The idea that our world is nothing but a computer simulation[145] has been raised many times, in writing and in movie adaptations.[146] In recent years, it has also received quite a bit of attention from bona fide scientists.[147] Below I present three arguments that support this claim. I happen to believe that we are not a simulation, and in several stories, for example, in "Existence is only a theory," I mentioned that the argument that we do not live in the Matrix is one that we generally accept although we cannot prove it. Nevertheless, thinking about the following three arguments has led me to the conclusion that perhaps the idea that we are part of a simulation shouldn't be dismissed out of hand.

---

[144] Conspiracy theories have a bad reputation, which is often justified. But because conspiracies do occur from time to time, some conspiracy theories are destined to be correct. I don't claim that the theory of life in the Matrix is correct. I claim that we should not reject it out of hand, and that discussing it may be useful.

[145] This story deals with a simulation that creates the laws of what we perceive as the physical world, and not with one that can interfere with their action and create "magical" phenomena like those described in the book *Sophie's World*.

[146] For example, the *Matrix* movies.

[147] I strongly recommend watching a video recorded during a discussion held by several senior scientists in memory of Asimov. It's a two-hour video, and I think it's worth every minute you devote to it.

8.6.1 *First argument: Quantization*

We know that many of the physical magnitudes exist in given portions referred to as "quanta," rather than continuously. We don't know what happens with other magnitudes (like space and time), but it is possible that they are not continuous either. Quantization is a feature we generally encounter in computer systems. Our usual perception of the world, however, tends to ignore it, and treats many magnitudes as continuous. If it turns out that all the magnitudes in nature come in fixed quanta, it will corroborate to some extent the claim that the behavior of the universe is controlled by a computerized system.

8.6.2 *Second argument: Limiting the speed of information transfer*

We know that the speed of light is limited and finite. According to the theory of relativity, all modes of information transfer between various points in the universe are similarly limited. Could it be that this limitation is due to the restricted computational power of the computer in which the simulation in which we live is being conducted?

8.6.3 *Third argument: We cannot obtain more effective problem-solving from nature than that of a computer[148]*

Understanding this argument requires a brief introduction to a field of knowledge known as "computational complexity." This field seeks to determine the extent to which computer algorithms are efficient in solving the problems for which they had been designed. The efficiency of an algorithm is determined by the order of magnitude of the ratio between the size of the input to the algorithm and the time or the number of elementary operations required to run the algorithm. For example, an order of

---

[148] However interesting, this argument may be wrong. Recent developments in quantum computing show that it may be possible to harness quantum effects to achieve more effective problem-solving with computers.

n operations is needed to find the maximum number in a set of n numbers.[149] The mathematical symbol for such complexity is (n), in other words, an order of magnitude of n. The number of operations needed to sort n numbers is bounded by $n\log(n)$ *multiplied by some algorithm-depended constant*, therefore, the complexity of the sorting algorithm is $(n\log(n))$.

Algorithms may be divided and grouped in several ways, but for the purposes of the present discussion, we are interested in the division between polynomial algorithms, whose complexity is limited by a polynomial,[150] and those whose complexity cannot be limited by a polynomial. The number of operations needed to execute a polynomial algorithm increases relatively modestly with increasing input size, whereas the number of operations required to execute a non-polynomial algorithm increases much faster with increasing input size, and the most powerful computers, even if many of them are working together, are limited in their ability to perform such algorithms when the size of the input exceeds several thousands.

Note that I am talking only about the algorithm (i.e., a certain way of solving the problem), and not about the problem itself. This is clear because the same problem may be solved in different ways, some of which are more efficient than others. There is a large set of problems that cannot be solved by any known polynomial algorithm. Among these, there is a whole set of problems that are interrelated in the sense that if a polynomial algorithm can be found to solve one of them, a polynomial solution can be derived from it for all the others. The problem of the traveling salesman[151] is a famous one in this set. "Not known" is not equivalent to "doesn't exist," and therefore one of the questions facing computer scientists is whether any polynomial algorithm exists for solving these problems. This is referred to as the P vs. NP problem,[152] and the assess-

---

[149] Whose number of digits is defined and known.

[150] Polynomial.

[151] The problem of the traveling salesman is as follows: "Given a list of n cities and the distance between any two of them, what is the shortest route that passes through each city exactly once and returns to the city from which it started?"

[152] P vs. NP problem.

ment of most people dealing with it is that the answer is negative, meaning that such algorithms do not exist.

I told this whole story because as far as I know, even resorting to nature has not helped us solve such problems.[153] Why can't we make nature efficiently solve problems that don't have an efficient computer solution? Doesn't this suggest that there is a computer system that dictates what is supposed to happen in it?

---

[153] Substantiating this claim requires going into details that do not necessarily interest all the readers, and therefore I include them in 11.3 Appendix to "8.6 Do We Live in the Matrix?"

Chapter 9

Concepts

This chapter includes stories in which I try to clarify my interpretation of various concepts used in philosophical discourse by comparing them, or by presenting and arguing my position with regard to them.

9.1 Is Science Just Another Religion?

Such claims are often voiced by the religious and the postmodernist polemicists who cannot handle the scientific evidence that contradicts their claims. The argument is used to disqualify or dwarf all scientific achievements and get rid of the "hecklers" once and for all.

The claim that science is just another religion is based on several misunderstandings, the first of which is a misunderstanding, of all things, of the term "religion." Nowadays, the term "religion" is used to describe rules of conduct that are not determined by legislatures, but that are alleged to originate from a source external to humanity (generally, God, but also other factors, as, for example, the principle of white supremacy). Clearly, this definition does not apply to science, if only because science does not deal with rules of conduct. It can furnish predictions about the possible consequences of behavior (such as, "if you smoke, you will increase your chances of getting cancer"), but it does not enjoin us to behave in one way or another.

This is not a trifling difference because the laws of behavior dictated by religions are often immoral and in many cases murderous. The misunderstanding of this difference is clearly expressed most thoroughly and pathetically in the words of "religious warriors," who seek to justify the events of the Holocaust by the theory of evolution rather than by the religion of Nazism.

A mitigated claim, which distinguishes between religion and faith, argues that science, like religion, is also based on beliefs, and therefore, one should not object to the beliefs that underpin religion. This argument is also mistaken. We all believe in certain things. We believe in the laws of reason, we believe that what our senses perceive represent in some way or another an objective reality, and so forth. These beliefs cannot be avoided, and if they are rejected, there is no point in trying to talk to anyone about anything because they are the common denominator that enables the conversation. Therefore, the religious also hold these beliefs.

The question is about additional beliefs, beyond these essential ones, where the religious believer differs from the rational person. The rational person is satisfied with the essential beliefs, and considers all the rest to be a collection of more or less confirmed theories. The believer adds to this a collection of beliefs that often contradict each other and reality.

This is a profound and significant difference, which also underlies the dynamism of science vs. the stagnation of religion. Religion has beliefs that supposedly are not of human origin, and therefore no one has the authority to change them.[154] The scientist, within the framework of scientific theories, works with hypotheses that are constantly confronted with reality, corrected, and at times scrapped as soon as findings contradict them.

An even more attenuated argument says something to the effect that "science has not proven its claims." Beyond the fact that the lack of proof of this or that scientific claim is no proof of the validity of religious claims (none of which have been proven and many of which are in outright conflict with reality or contradict each other), this argument suffers from a misunderstanding of the scientific enterprise. Scientific theories are never

---

[154] Some minor changes are made occasionally, through revised interpretations of the text, but the main core remains the same, and even the most revolutionary reformer never claims that the initial holy scriptures are wrong.

proven. If it were possible to prove everything, there would be no need for science, and mathematics would be sufficient. Science exists because of the recognition that it is impossible to prove any non-trivial claim that is not mathematical, and mathematics does not investigate the world because it deals with all possible worlds. In science, mathematics is used to formulate theories and derive predictions from them, but science itself relies on the findings of experiments. Any theory, however beautiful and mathematical it may be, collapses if its predictions don't match the findings of experiments and observations, in stark contrast to religious convictions that survive all refutation.

As a mathematical structure, a scientific theory is based on a collection of axioms of mathematics, a collection of experimental findings, and hypotheses derived from the results of experiments. Based on these, it seeks to predict, using mathematical calculations, what will happen in future experiments. Of course, the core of the theory lies neither in the findings nor in the process of deriving conclusions from them, but in the hypotheses that try to pull together all the findings without any contradiction. In principle, the collection of the findings can be "packaged" in several ways (some of which may not be known at the time of the formulation of the theory), and therefore the theory always remains a hypothesis. Our confidence in the correctness of the theory increases as it successfully withstands more and more experiments and observations, but a single experiment that does not match its predictions is sufficient to indicate that it is not a full description of reality. As a result, we seek to limit the areas of its applicability in its current formulation, and try to either expand or to replace it, to increase once again the scope of its applicability. For example, when Newton wrote his book *Mathematical Principles of Natural Philosophy*, the rules of motion he described were an excellent explanation for all known results of experiments and observations. Einstein's theory of general relativity was another way to generalize the same observations, but it was not known at the time when Newton formulated his theory.

Newton's theory remained a hypothesis, but was eventually replaced by Einstein's theory, which did better in predicting or explaining additional observations that were not known in Newton's days.

9.2 Logic and Mathematics vs. Science: What They Have in Common and What Separates Them?

9.2.1 *What is science?*

The word "science" serves us in two ways. One meaning is "the scientific method" or "a method for exploring the world and exposing the phenomena that are true in it, or at least describing them as accurately as possible." The second meaning is "a collection of claims about reality obtained to date as a result of the use of the scientific method." The scientific method is based on examining the suggested theories as a possible description of some aspect of reality, by verifying them against reality.

A theory is a collection of basic assumptions (what is commonly referred to in mathematics as "axioms"). When formulating a theory, one *does not* undertake to prove it logically, or to develop logically all the conclusions arising from these assumptions. Verifying the theory against reality is based on a comparison of the claims of the theory (its basic assumptions and the assumptions obtained logically or mathematically from the basic assumptions) with what happens in the world. For a theory to be considered scientific, such verification should be possible. Science has no interest in theories that do not correlate with the observable reality because such theories have no effect on our lives and it is difficult to claim that they have any meaning.

A scientific theory may be wrong. This does not rule out its still being scientific. On the contrary, the fact that we could prove that some theory is wrong shows that the theory is scientific: the refutation of the theory is obtained through verification against reality, and the possibility of such a verification is an essential demand we place on a scientific theory.

But a scientific theory that was proven wrong, despite its scientific nature, is not included in the body of knowledge obtained through the scientific method, and therefore it is not included in the word "science" in the second sense I mentioned (a collection of claims about reality obtained to date as a result of the use of the scientific method). This collection contains only scientific theories that have not yet been refuted, although there is a possibility in principle to disprove them if they do not correctly describe reality. The longer the accumulated list of failed attempts to refute a theory, the greater our faith in its correctness.

It is important to understand that the heart of a theory lies in its basic assumptions, and that all the rest are conclusions derived logically from these assumptions. Therefore, it makes more sense to be convinced of the correctness of a theory whose basic assumptions are easy to test empirically, because all the rest is logic and mathematics. A theory whose basic assumptions are successfully verified against reality is more persuasive than a theory whose conclusions have been successfully verified against reality because it may be possible to reach those conclusions also on the strength of various other basic assumptions. When Newton's laws of motion were formulated, all known observations about the motion of rigid bodies could be explained. The same observations could also be explained by Einstein's relativity. Further observations were required to enable scientists to conclude that Einstein's laws were more accurate.

Sometimes it is not possible to test the basic assumptions directly. No one expected to see an electron when its existence had been suggested. There was simply enough evidence to support its existence in the results of experiments that were obtained after the theory positing the existence of electrons predicted them. But if the assumption can be observed directly, it is easier to accept the theory.

Theories often describe a chaotic reality, in which many factors play a role. Therefore, a mathematical solution cannot produce reliable predictions for the simple reason that not all the data that must be taken into account are available. Predictions provided by such theories are fuzzier, but if the basic assumptions can be checked directly, there is no doubt that the theory is essentially correct. Such is the case of evolution. From this point of view, evolution has a great advantage over, say, the dark matter theory, where all the evidence is indirect, that is, the evidence verifies the conclusions of the theory that posits the existence of dark matter, not its basic assumptions. Nevertheless, the dark matter theory is quite readily accepted by a large portion of the scientific community. The basic assumptions of general relativity theory are also difficult to observe directly, and the evidence we have for its correctness comes mainly from the fulfillment of its predictions; this is less powerful evidence than the direct evidence of the fulfillment of the basic assumptions of the theory of evolution.

Even if the theory cannot predict the future accurately or in every aspect, it can sometimes predict connections between different pieces of data to be obtained in the future. Evolution, for example, cannot predict what creatures will develop because mutations occur randomly (at least as far as our ability to discover their causes is concerned), but this doesn't mean that it cannot predict anything. It can predict, among others, connections between evolutionary trees obtained from the analysis of various genes, even if it cannot say in advance which trees will be obtained. This is a bit reminiscent of our ability to know that the spins of two quantum-entangled particles will be correlated when we measure them, without being able to know in advance what these spins will be.

Although people often repeat the argument that evolution does not make predictions, a claim that happens to be clearly incorrect, even without predictions, there can be no doubt about the correctness of evolution because its basic assumptions (the existence of replicators and the existence of competition for resources, the success of which affects the degree of replication) can be directly verified.

The situation is much more problematic when we make basic assumptions that do not hold in reality. As long as there is no way to determine at least in what proportion of the cases the assumptions hold, or to define an upper bound for the error that may be obtained as a result of the difference between the situation that the theory assumes and the situation in reality, all we have is a model whose connection to reality is obscure.

In the past, an unquestioned assumption was that the space in our universe was Euclidean. Today, we know that it is not. In other words, the basic assumption that physical space is Euclidean has been refuted, and today we know that space is curved, and that its curvature is determined by mass and energy. Nevertheless, we can still use Euclidean geometry in many cases because we have a way of calculating the difference between the results obtained using it and the more accurate results. We can estimate this difference because we have another theory, which has not yet been refuted by measurements. Therefore, we calculate the distance between Euclidean geometry and reality simply by calculating the distance between it and the geometry of general relativity. Thus, the theory of Euclidean geometry of space, although not a correct scientific theory, is useful.

9.2.2 *How do logic and mathematics fit into this story?*

Logic and mathematics are the ways in which we calculate the results derived from the axioms upon which we based the theory. Mathematics, although it is considered the "queen of sciences," is not science in the accepted sense of the word because it is *a priori,* in other words, it is a theory that is not derived from experience (as opposed to science, which is based entirely on experience).

It is customary to say that while science explores the world in which we live (and makes use of logic and mathematics), logic and mathematics themselves explore all possible worlds, even those that are not the one in which we live. Indeed, the discovery of the fact that the geometry of the world is not Euclidean did not strike down the correctness of Euclidean geometry as a mathematical structure, or the possibility of continuing to study this mathematical structure.

This is a useful and convenient view of logic and mathematics. It is the generally accepted view, and I don't recommend changing it, although it is not entirely accurate. Indeed, logic, and consequently mathematics, is a scientific theory for all intents and purposes, based on experience like any other scientific theory. However, unlike conventional scientific theories, this experience is not the product of human beings or any other conscious creature but of evolution. We are born with built-in "logic circuits" within our brains, and these circuits are the result of a very long series of experiments on animals, in which creatures whose logic reflected reality better have enjoyed an advantage. Therefore, mathematics is part of science in the second sense I mentioned (a collection of claims about reality obtained by using the scientific method, but in this case, by evolution, not by people).

But that's just between us. It has no practical use, because we cannot even imagine a reality in which logic does not work, therefore, it is more convenient and useful to continue to view it as holding *a priori,* without depending on experience, because we are unable even to imagine worlds where logic does not hold. I believe that there are no worlds of this type, but can I, a product of evolution, be expected to think otherwise?

9.3 Existence is Only a Theory

I'm a realist, in the sense that I believe in an objective reality. But when I know that something exists, I almost never know it directly, and almost always I postulate a theory about the nature and characteristics of that reality.

What does it mean that something "exists?" Does the sun exist? Do electrons exist? Is evolution a fact? Does God exist? We seldom dwell on these questions, and yet some of us provide different answers to them, although these appear to be objective questions, and the answers of different people concern the same reality. The source of the difference between the answers lies not in the common reality, but in the fact that people interpret differently the information collected by their senses. Indeed, all our arguments about the existence of things reflect the way we interpret the information collected by our senses.[155]

The claim that existence is only a theory may sound unreasonable with regard to the sun, after all, we see it with our own eyes. (I use this example because the sun represents something we have no doubt about.) But the sun is merely an object whose existence we infer from the pattern of photons impinging on our retinas. The process of drawing conclusions about the existence of the sun is unconscious and automatic: anyone whose eyes are working believes that there is a sun, and no one interprets the pattern of photons on their retina differently. This interpretation of visible information is also reflected in the optical illusions we experience, which we cannot free ourselves from, even if we know they occur.

The same claim sounds more reasonable with regard to the electron. Has anyone ever seen an electron? If not, why did we come to the conclusion that it exists? Quite simply, because by assuming its existence we can explain many phenomena that we do experience with our senses, and

---

[155] This is true also of things in whose existence we believe because we believe what others have reported about their existence, or about the input of their senses. But for simplicity's sake, I focus here on things whose existence we infer without the mediation of others.

without contradicting any of our sensations. The existence of the electron is actually a theory. The existence of the sun is also a theory — and obviously it cannot be proven, because we cannot even prove that we are not living in a Matrix.[156]

As far as you are concerned, my existence is also a theory: to most of you, it explains how the text you are reading was created; to some of you, it explains how my reflection is formed on your retinas in certain circumstances, or how the air vibrations that cause a sense of voice recognition are created (a theory in itself). So is evolution merely a theory? Contrary to the usual claim of the believers in the existence of evolution, I argue that it is. Evolution is only a theory. But because everything we think we know is theory, the word *only* is superfluous.

Finally, is the existence of God a theory? The answer is, "Yes, it is a theory, but a theory disproved for every sufficiently clear definition of God."

9.4 Why Am I a Realist?

I see *realism* as a worldview according to which there is a reality outside myself. This is contrary to what *solipsism* claims, which is that only my consciousness exists with absolute certainty. Why do I accept realism and reject solipsism?

I believe that it is impossible to answer almost any question about reality with definite certainty. It may be that we live inside the Matrix

---

[156] Some argue that if everything is mediated by our senses and our intellect, there is no justification for using the term "reality." They're wrong. There are certainly things that we know directly to be part of reality: these are the things that occur within ourselves, inside. Descartes said, "I think, therefore I am." The facts that we exist, that we feel certain feelings, and think certain thoughts are all part of the reality that is unchallenged, even if this reality is not directly accessible to others — in case they happen to exist... Indeed, all these facts remain true even if we live in a Matrix.

Note that this minimalist approach does not negate the relevance of science. According to this approach, one of the roles of science is to correctly predict what we will feel under these or other circumstances, even if our feelings have no connection with an objective reality outside of us. Scientific theory is refuted when it predicts feelings different from those we feel in practice.

(I don't think so, but I cannot disprove it), or that all other people exist only in our imagination (I don't think this is the case, but I cannot conclusively prove that it is not), and so on. Nevertheless, I still think that it is possible to assert compellingly that there is a reality that is external to our consciousness. The reason for my confidence lies in the correspondence between some of my observations[157]:

> When I see a stone striking my body, I also feel the impact.
>
> When I don't see the stone but I feel the impact on my back, I can, if I turn quickly, catch a glimpse of it as it falls away from my body.
>
> When I hear the voice of a certain person and turn my eyes to the direction of the sound, I usually see that same person speaking.
>
> When I smell something noxious in the kitchen and look hard enough, I find the rotten potato or onion in the drawer.

There is nothing in my consciousness that creates these correspondences, therefore, there must be something outside of it that causes them. In other words, even if all of my sensory experiences do reflect something other than the true reality, the correspondence between them attests to the existence of such a reality. It may be possible to claim that the source of the correspondence is the subconscious, but there is no contradiction in my claim because the subconscious, as its name suggests, is not part of my consciousness.[158]

---

[157] Schrödinger wrote in a letter to Einstein: "The conception of a world that really exists is based on there being a far-reaching common experience of many individuals, in fact of all individuals who come into the same or similar situation with respect to the object concerned" (https://www.sbhc.org.br/arquivo/download?ID_ARQUIVO=1102). I don't find this argument convincing because whoever assumes the existence of other people already presupposes the existence of an external world.

[158] Some kind of subconscious is necessary in the first place, so that our consciousness can encounter things that were not known to it in advance, so that the word "observations" will have any meaning, but the correspondence between the various observations requires the existence of a much more sophisticated subconscious that can retrieve correlated observations. Indeed, this is a subconscious that operates like the Matrix (a term borrowed from a movie by this name, where the input provided to the brains of people is generated by a computer and not by their senses).

Admittedly, I also subscribe to scientific realism (according to which the inputs from our senses represent approximately the external world as it is), but in contrast to philosophical realism (that only asserts the existence of an external world), which in my opinion is a logical necessity, scientific realism is a working assumption the truth of which I cannot prove.

9.5 Beyond the Reach of Science

I often hear from clerics or new age people that this or that topic is beyond the reach of science.

Such statements are generally used to defend theories that people try to promote in the teeth of scientific or rational refutation. The wording of the statement varies according to context, but it is always meant to throw dust in our eyes. Below are a few examples of sentences I heard from such people, and my answers to them.

First example: The article attempted to use rational means to rule out irrational phenomena.

My response: This sentence mixes apples and oranges. Rational thinking is a way of examining phenomena and reaching conclusions about them. The phenomena themselves are not rational or irrational; they either exist or they don't. Rational thinking helps us judge whether or not they exist.

Second example: It must be remembered that a rational approach and scientific research are limited by the physical dimensions we live in.

My response: Correct. We have no way of knowing anything else, so people who talk about things they claim exist in other dimensions, talk about something they have never observed (because they are limited by the physical dimensions they live in). They can claim with the same degree of justification (zero) that the flying spaghetti monster and the tooth fairy exist.

Third example: Since metaphysical phenomena are irrational, to prove or refute their existence one must know how to use irrational means.

My response: Beyond my argument about mixing apples and oranges, I also note that we have no thinking tools available that can prove or refute phenomena by irrational means. Even claims raised within the framework of metaphysics must not contradict facts known to us; the role of metaphysics is to help us explain facts, not ignore them. For example, the claim about the existence of the soul as a separate entity from the physical body contradicts the findings I quoted in "Chapter 1. Soul, Mind, Will, and Self (Are) Matter." This contradiction does not disappear by defining the mind as a metaphysical entity.

In conclusion: There is no point in making claims about things that science cannot investigate, because everything that affects our lives can be scientifically investigated. Sometimes several scientific theories are mathematically equivalent, which means that anything predicted by one of them is also predicted by the others. This is, for example, the case of the different interpretations of quantum theory. The decision between these theories is not possible in principle, but there is a rule of thumb that scientists follow whenever possible: it is called Occam's razor, and it recommends "choosing the simplest explanation that involves the least number of concepts and rules."

For example, to explain gravity, it is possible to present the following theories:

— Newton's theory of gravity
— General relativity
— God drives mass and energy along paths that are consistent with general relativity

The first is rejected because its predictions do not match the findings, but the last two are mathematically equivalent. Occam's razor suggests that the first of the two is preferable.

9.6 Atheist or Agnostic?

The agnostics claim that it is impossible to know with certainty whether God exists. I say that it is impossible to know with certainty almost anything, including whether or not God exists. So why do I still call myself an atheist?

I do this for several reasons, the most important of which are:

1. I distinguish between "knowing with certainty" and just "knowing."
2. The God of agnostic philosophers is not the one whom religious believers are trying to sell us, but only an impersonator by means of which they wish to sell us another God.

Agnosticism is defined as ignorance of the answer to the question whether God exists, and not as "lack of irrefutable knowledge." The term "know" does great service, usually without the need to add the words "with certainty." When someone takes the trouble to use the phrase "know with certainty," he indicates that he knows (even if he may not be aware of it) that this expression means something different, otherwise he wouldn't bother to unnecessarily complicate the statement.

I believe that the word "knowledge" is used to describe a state of mind in which we attribute a very high degree of probability to something. I say that I know they call me Michael, but I don't know this for certain because I don't know for certain that anyone is calling me at all. I say that I know that the sun is shining, but I have no evidence of it other than my sensory input, and I don't know for sure whether this input gives me a reliable account of reality; perhaps I live in the Matrix.

When people talk of "certain knowledge" of a particular claim, they usually associate it with the possibility of proving the claim. I don't know with certainty that there is no God because the claim "There is a God," as well as the claim "There is no God," cannot be proven. In the absence of any proof, is it justified to claim that I don't know whether there is a God? Not if it is not justified for me to claim that I don't know what my name is. People tend to equate knowing something with the ability to prove it. This not only obviates words in the language, but it also leads to genuine paradoxes, as I show in "10.3 The Pop Quiz Paradox."

To sum up this point, because to the best of my knowledge there is no evidence of the existence of God, I attribute a very high degree of probability to the claim that God does not exist, and therefore, it is justified for me to say that I know there is no God, or in other words, that I'm an atheist. If I described myself as an agnostic, I would lead some readers to conclude that I attribute a similar probability to the possibility that God exists and to the possibility that God does not exist.[159]

In the present context, the next point is even more important to me than the divagations on the meaning of the word "knowledge." The word "God," as it appears in various religions, has a much more concrete meaning than what is accepted in philosophical discourse. By using the word "God," philosophers fall into the trap that religious believers (or they themselves) have set, because religious believers find it much easier to defend the existence of the philosophers' undefined God than the existence of the God that their religion describes. Therefore, they always divert the debate towards the existence of an undefined philosophical God and away from the debate over the existence of the God of religion.

Consider, for example, the God of Judaism, who is definitely not the one about whom Einstein spoke when he said that God does not play dice.[160] The omnipotent and omniscient God of Judaism created the world and determined the content of the Torah. The existence of such a God is easy to refute, for it is inconceivable that such a God wouldn't know that the hare he created doesn't chew the cud (Leviticus 11:6), or that the Tigris and the Euphrates that he created do not originate from a common source (Genesis 2: 10–14). Bear in mind that the religious description of God is intended to justify His authority in determining the laws of behavior dictated by religion. This in itself justifies the use of the word "atheist," which emphasizes the fact that the person described by this term does not accept this authority.

---

[159] The expression "agnostic atheism" describes the situation I refer to as "atheism," but because lack of certainty is characteristic of everything we call "knowledge," there is no need for such a term.

[160] Einstein gradually understood the mistake he had made when he used the word "God," and expressed this in increasingly pointed remarks during his life. His most explicit comment was "The word God is for me nothing but the expression of and product of human weaknesses, the Bible a collection of venerable but still rather primitive legends. No interpretation, no matter how subtle, can (for me) change anything about this." https://www.nytimes.com/2018/12/02/nyregion/einstein-god-letter-auction.html.

Supporters of religion tend to ignore these comments and claim that Einstein believed in God.

9.7 Beyond Doubt

The road from Hell is paved with good questions.

One of the things that distinguishes an atheist from a religious person is the understanding that blind faith is a dangerous strategy, whereas doubting is a habit worth adopting. Many people identify doubting with scientific thinking. In my opinion, these people miss a large part of the picture. Doubting is indeed necessary for formulating a proper understanding of the world, but it is not enough. What we need, beyond doubting, is to know to ask.

Doubting is also a form of questioning. It's always the same question: "Is what I think I know really true?" But the questions that interest us are:

1. What is true?
2. Why is truth such and not otherwise?

The vast majority of people don't ask anything. Few of them, those referred to as "critical thinkers," ask the question "Is what I know the truth?" Fewer ask the question "What is true?" Only a handful ask "Why is truth such and not otherwise?"

The people who ask the last two questions are usually philosophers and scientists, but I suggest that everyone should ask them because engaging with them greatly enriches life. Take, for example, the following questions[161]:

1. Why, when folding a piece of paper, the fold is always a straight line? We all know this is the case; but have you asked yourself why?
2. What is the force driving the tides? Everyone knows that they are related to the power of attraction of another body (on Earth, especially of the Moon); but how?
3. Why does the Moon always face the Earth with the same side? Most people know the term "the dark side of the moon," which refers to the side we never see from here, and many people know that the

---

[161] In the spirit of the Passover *Haggadah*, according to which, "he who doesn't know how to ask, you initiate it for him."

configurations we see on the Moon (which describe its geography) are always in the same formation, but most people don't ask why this is the way things are. How does the Moon's revolution around the Earth coordinate so precisely with its rotation?

4. How do scales operate? When the weights on both sides are equal, why do the scales balance and not just stay where they were in the first place? Why, when one side is heavier, it doesn't completely descend to the bottom but stabilizes in a position that expresses the difference between the weights?

5. How do the movements we perform on a rope swing increase the motion of the swing?

6. How long does it take the Earth to complete one rotation around its axis?

7. In school, we usually learn two "physics." In the first one, a point of mass reaches equilibrium when the sum of the forces acting upon it is zero; in the second, it reaches equilibrium when the potential energy is minimal. Why can we count on the assumption that the calculations carried out on the two systems will yield the same results?

In the above list, I attach special importance to the last question because it touches upon a subject that is taught in schools (in Israel), and in most cases the teachers do not draw the students' attention even to the fact that this is a question that needs to be asked, let alone answered.

I consider this question to be important also because of its ability to exemplify the fact that avoiding asking the questions I mentioned at the beginning (specifically, "Why is truth such and not otherwise?") exposes all of us to brainwashing. In this case, our minds are washed with a true claim, but in many other cases, especially if we fall victim to religious education, we will learn as "truth" false claims as well.

I invite you to:

1. Give an account of yourself on the topic of "Have I asked myself questions of this type? (and if not, why not?)."

2. Read the stories "Profession" and "Jokester" by Asimov, which also emphasize the importance of raising questions.

Part 2

Not So Stories

Chapter 10

Paradoxes

One of the goals I set for myself in creating the blog is to improve people's resilience to lies by increasing their awareness of the power of critical thinking and of the power of reason. The stories on paradoxes were written mainly for this purpose. I see paradoxes as a first-class educational tool, because unlike solutions to ordinary problems, which take us from a state of "not knowing" to one of "knowing," the resolution of a paradox takes us from the state of "knowing the wrong thing" to one of "knowing the right thing."

10.1 A Paradox is Always an Error in Judgment

A paradox is a situation in which two seemingly legitimate ways of thinking lead to contradictory conclusions.

The solution to a paradox is not achieved by preferring one way of thinking (perhaps the one in which the solution seems more reasonable) over the other, but by identifying a mistake in at least one of the alternatives (perhaps in the one in which the solution appears unreasonable). Any other solution fails to reconcile the contradiction. This is why paradoxes are important: their solution forces us to delve into fundamental problems in our thinking and solve them.

Below I use the *Achilles and the turtle* paradox to explain briefly what I mean. This paradox was formulated by the Greek philosopher Zeno more or less as follows: Achilles runs at a speed of 10 meters per second,

10 times faster than the turtle, so he decides to grant the turtle a head start of 100 meters. It is clear (by solving the linear equation) that Achilles will overtake the turtle within a short time, but according to Zeno, under the terms of the race, Achilles will never catch up with the turtle. The reason for this is that while Achilles runs the first 100 meters and reaches the turtle's starting point, the turtle will have advanced by another 10 meters, and therefore is still ahead of Achilles. As Achilles continues and runs the additional 10 meters, the turtle will have advanced yet another meter, and again is ahead of Achilles, and so on. By the time Achilles reaches each point where the turtle had been first, the turtle will have moved to a point further ahead. Therefore, Achilles will approach the turtle, but will never be able to catch up with it.

Where is the mistake?

We reach the conclusion that Achilles will never catch up with the turtle through a description of an infinite series of events in which Achilles has not yet caught up with the turtle. This series indeed exists in reality, and can also be reflected by describing the movements of Achilles and of the turtle through regular linear equations that represent motion. But it is precisely at this point where the error in judgment takes place, when based on the existence of that infinite series of events we conclude that Achilles will never catch up with the turtle. In reality, all the events in question, although infinite in number, take place before a certain time (the series of smaller and smaller time intervals separating them converges), and therefore it is incorrect to infer from their existence anything about what happens after this time.

In other words, it was right to conclude from the existence of the series of events that Achilles would not catch up with the turtle before, say, 10:00 a.m., but it is not right to conclude that he will never catch up. The error is one of over-generalization, and can be compared to the following one: I walk around my house and ask all the people I meet what is their country of residence. Everyone answers "Israel," and I conclude that all the people in the world live in Israel.

Note that we don't always seek to avoid baseless generalizations. Sometimes there is justification for not limiting our generalizations. If, for example, we observe the scientific enterprise, we find that it constantly commits what I have just described as an error, because all our experiments have been conducted in the past, and our conclusions about the laws derived from them concern the future as well. We allow ourselves to do so because we believe in the validity of some axiom in relation to the laws of nature, according to which these laws do not change over time. In the case of Zeno's problem, of course, we naturally do not rely on an axiom of this type regarding the position of Achilles and the turtle because the movement is defined by <u>changing</u> their positions as time progresses.

10.2 I Am Who I Am: Paradoxes of Self-reference

Discussions about religion and faith often lead, deliberately or unconsciously, to problems with self-referential claims. The main feature of such problems is that self-reference is a logical error leading to many paradoxes.

The most familiar example of paradoxes of self-reference is the so-called "liar paradox." Its short and clear illustration is the phrase: "this sentence is a lie." If we assume that the sentence makes a true claim, we can conclude that it makes a false claim, and *vice versa*, if we assume that it makes a false claim, we can conclude that its claim is true.

Below I address some of the more complex examples of the paradox, and the reason why self-reference creates paradoxes, but first I would like to bring examples of this paradox as they appear in the daily discourse of people of religious and new age persuasion. A common version of the paradox is, "There's no absolute truth," or in a slightly different form, "There's no **one** truth." This argument is frequently voiced in the spirit of the new age people to belittle the efforts and achievements of science in exposing the truth. When I come across postmodernists who raise this argument, I tend to ask, "Wait, is this argument that you just made not an absolute truth either?"

Here is another one: "Every creature has a creator." This is a typical line used by religious proselytizers, taking advantage of the common root of the words "creature" and "creator," but falling into another instance of the same paradox. They claim that every creature has a creator, wishing to release God from this requirement. But what exactly is it that makes God "not a creature," and why was there no need for a creator to create Him?

And again: "Every law has a legislator." I have heard this claim also from religious proselytizers, who wanted to show that even if nature did follow certain laws, God was needed to enact these laws. My answer is usually based on "And who exactly enacted this law, according to which every law has a legislator?" The intention of this question, which I usually take the trouble to explain, is that their claim obliges those who accept its conclusion to assume the conclusion, which in this case is to assume that there is a God even before the question of its existence is being raised.

It is easy to fall into the trap of contradictory self-references, and, as the above examples show, there are people who specialize in it. Even mathematicians occasionally do it. But because the mathematics environment does not tolerate contradictions, these lapses have led to a more thorough investigation of the reasons for the contradiction and to the understanding that caution must be exercised when formulating claims and definitions that contain self-references.

The problem that made mathematicians aware of this issue appears to have been Russell's paradox. Russell assumed the existence of a set containing as members all the sets that are not members of themselves, and asked whether such a set is a member of itself or not. The paradox is created because if the set is not a member of itself, then it is one of the sets that are not members of themselves, and therefore it must be a member of itself. The opposite is also true, because if it is a member of itself, it is not one of the sets that are not members of themselves, and therefore is not a member in the set of all sets that are not members of themselves, that is, it is not a member of itself.

This paradox threw the mathematical world into a tizzy, from which it escaped through various manifestations of one simple insight: it is not possible to define something based on itself. Anything on which a definition is based, must be defined first.

If you think about it, this is pretty obvious: if we want to define something, we cannot rely in this definition on something that is not defined, and at the moment of its definition, the term we want to introduce has not yet been defined. Therefore, among others, talking about a set that is not a member of itself is asking for trouble because the set is defined by the collection of the members in it, and as long as not all the sets have been defined (including the set in question), it is not possible to verify which of them are members of themselves, so that in order to be able to define the set in question, we must first define it.

Self-reference does not have to create a paradox, but even when no paradox is created, its meaning is problematic. Consider the opposite of the liar's paradox: "This sentence is true." This is a sentence that simply has no meaning, and in practice, it is also impossible to determine whether it is true or false: for if it is true, it is true, and if it is false, it is false. Both assumptions are equally acceptable.

Self-reference is the source of many paradoxes that philosophers have long struggled with. Note that some of what I write here may be unacceptable to many philosophers, but I'm convinced that I'm right and that someone who thinks otherwise is wrong. At times people tell me that I express in categorical terms things about which others disagree. I could introduce each statement with the expression "in my opinion," but when I'm fairly sure that my opinion is correct, I avoid doing this because it makes communication cumbersome. After all, it is clear that what I write is my opinion, and even if I write a sentence such as, "In Einstein's opinion there is no God," the meaning of the sentence for the reader is that my opinion is that Einstein believes there is no God.

I once witnessed a discussion conducted by Amnon Lipkin Shahak. One of the participants, on his turn to speak, began by: "*I* say…" Lipkin Shahak interrupted him: "What you say is always what you say!" He was a wise man. Naturally, these statements ("in my opinion…" "I say…" etc.) are examples of self-reference, and therefore, in many cases they do not carry significant information.

Many famous paradoxes are classified as self-reference paradoxes. One of the paradoxes described in this book, in "10.3 The Pop Quiz Paradox," also has to do with self-reference. Finally, there are paradoxes that are not based on self-references, such as the Raven paradox, described in the story "10.4 Unpredictably Rational: All Ravens Are Black," the paradox of Zeno, described in "10.1 A Paradox is Always an Error in Judgment," and others.[162]

---

[162] A list that attempts to classify paradoxes by their type can be found in the Wikipedia entry for the term "Paradox."

I would also like to mention briefly that paradoxes of self-reference can be useful. For example, the proof of Gödel's incompleteness theorem is based on the coding of logical propositions within natural numbers, and the construction of a number whose translation into the world of logic is something like "this theorem cannot be proven." The translation into the world of natural numbers makes it possible to evade the self-reference. (This is not a mathematical text, so the above description is not complete or even accurate, it is merely intended to be concise and not technical.)

Those dealing with mathematics and computers often encounter recursive definitions. Definitions of this type do not reflect self-reference, but a detailed discussion would unnecessarily confuse the present story. I presume that those who know what a recursive definition is, understand why this is not a case of self-reference.

Occasionally, at nostalgic meetings, my friend, Prof. Arnon Avron mentions another beneficial use that I found for self-reference. He tells the story that when he once approached me and asked whether he could ask me a question, I replied, "You had a chance to ask one question, and you just wasted it."

10.3 The Pop Quiz Paradox

In the story on paradoxes of self-reference, I mentioned the pop quiz paradox. This complex example deserves a separate discussion. Here it is.

The paradox is known in different versions[163] and can be illustrated by the following story.

A teacher informs students that at 9 a.m. of one of the days in the following week they will have a pop quiz, and that the students will have no way of knowing on which day it will take place until the quiz begins. Trying to figure out when the pop quiz is going to be held, the students performed the following analysis:

---

[163] The most common one is the unexpected hanging paradox. Structurally, all the versions are identical.

- Surely, the quiz cannot be held on Friday, because if it is not held by Thursday, we will know for sure at the end of the day that it will take place on Friday, which contradicts the teacher's statement that until the quiz begins we won't be able to know on which day it will take place.
- Therefore, Thursday is the last day on which the quiz can take place. But if so, the quiz cannot be held on Thursday, because if it does not take place by Wednesday, we will know for sure at the end of the day that the quiz will take place on Thursday, contrary to the teacher's statement that we will not be able to know on which day the quiz will take place until it begins.
- Wednesday, Tuesday, and Monday were also disqualified in a similar way.

The students reached the conclusion that, based on the conditions determined by the teacher, the quiz cannot take place, so therefore didn't bother to prepare for it. Imagine how great their surprise was when, on Tuesday, the teacher walked into the classroom and announced the start of the promised pop quiz. Where was the error in the analysis conducted by the students?

Various explanations have been offered for the paradox, but I haven't yet found one that satisfies me. My resolution of the paradox relies on the difference between the meaning of the word "know" and the meaning of the phrase "can prove." In this case, the difference between knowability and provability originates in the fact that, based on a system of axioms that includes an internal contradiction, it is possible to prove any claim[164] (and its negation).

It is clear, therefore, that when we prove a claim based on a system of axioms that involves an internal contradiction, although we were able to

---

[164] A method of proof that works for every proposition is that of proof by contradiction. In this type of proof, we try to find a contradiction in the set of propositions, which we can prove when we add to the axioms the negation of the proposition we want to prove. If our set of axioms is inconsistent and we want to prove proposition P, we can adjoin the proposition "P is false" to the set of axioms, observe a contradiction (which is certainly there because the set of axioms is inconsistent), deduce that "P is false" is false, and that hence P is true.

prove that the claim is indeed derived from the axioms, we don't know that it is true because we know (or at least, can know) that the system of axioms is inconsistent and can also be used to prove the negation of the claim. The system of axioms that the students have gleaned from the teacher's statement includes an internal contradiction. It is easy to identify this contradiction when we talk about a single day rather than a week. Consider someone stating: "Tomorrow I'll give you a quiz, and you won't know on which day it will take place." In this case, everyone can see that the teacher contradicts himself. When it comes to several days, however, the formulation of the teacher's statement becomes convoluted, but the contradiction doesn't disappear.

We reach the conclusion that a certain set of assumptions is not consistent when it is possible to prove based on it something and its negation, which is exactly what is happening here. We can prove that the quiz will take place on one of the days (one of the axioms is that a quiz will be held this week), but we can also prove the opposite, that the quiz will not take place on any of the days (which is what the students did). In other words, if the term "know" is interpreted to mean "possibility to prove," the teacher imposes on the students a system of axioms that includes an internal contradiction, and therefore they can use it to prove a claim and its negation.

To better understand this, let's look at a similar problem where, instead of "you won't know," the teacher says, "you won't be able to prove." In this case, it is clear that the teacher actually makes a promise that he doesn't keep, because for each day the students can prove that the quiz will take place on that day,[165] as well as that it will not. This shows that the problem begins with our interpretation of the term "knowing." Once we replace this term in the formulation of the problem with the meaning the students used in formulating their considerations (the meaning of "being able to prove"), everything becomes clear because we get an inconsistent set of axioms that enables us to **prove** everything, but doesn't enable us to **know** anything.

As in many other paradoxes, it is possible to associate the internal contradiction in the system of axioms with the self-reference it contains,

---

[165] Remember that we've seen that the set of axioms is inconsistent and therefore it enables us to prove any claim.

when one of the axioms maintains that there is a conclusion that cannot be deduced from the combination of itself and the other axioms. The teacher can keep his promise in the way in which he formulated the problem, but the students translate his promise into axioms ignoring both the cases in which "knowing" and "being able to prove" are not equivalent, and the self-reference leading to internal contradiction. Because it is not possible to rely on a system of axioms with an internal contradiction to "know" something, the students really won't know when the exam will take place.

10.4 Unpredictably Rational: All Ravens Are Black

But can something *not black* serve as evidence for the claim that all ravens are black? Logically equivalent arguments are not always equivalent when data about the real world is taken into account.

The phrase "all ravens are black" is logically equivalent to the phrase "everything that is not black is not a raven." Think about this for a moment, and make sure you agree with this statement before you continue reading, because there is no point in reading on before this equivalence is understood.

Now, imagine that someone is trying to convince you that this statement is correct. To do so, he can take you on a worldwide tour and show you ravens everywhere, so that after having seen many ravens and ascertained that they are all black, you accept the claim. But, he can also follow a different strategy. He can take you on a tour around the globe and show you many things that are not black, pointing out to you that none of these things are ravens. Will you find this convincing? *Nevermore.* Why wouldn't a reasonable person accept the results of such an experiment as confirmation of the claim "all ravens are black?" After all, the second experiment is aimed at confirming the claim that "everything that is not black is not a raven," which is equivalent to the claim that "all ravens are black."

Does our intuition cause us to reject affirmations of logically equivalent arguments only because of their formulation? I don't think so. On the contrary, what is happening here is that sometimes our intuition is more intelligent than the conscious analysis we carry out in our decision-making process. Our intuition is correct in rejecting the confirmation of "everything that is not black is not a raven," and there is a reason for doing so. Both experiments are designed to disprove the hypothesis. In both cases, the more instances we accumulate in which the hypothesis is not disproved, the greater our belief in its correctness. And in both cases, the hypothesis is disproved by a single contradictory finding, which in both cases is the same: a raven that is not black (in other words, something that is not black but is a raven).

Because the number of ravens is far, far, far smaller than the number of things that are not black, the group of non-black ravens makes up a much, much, much smaller percentage of the group of non-black objects than of the group of ravens. Therefore, the chances of an observation to refute the hypothesis are much higher if we check the ravens than if we check the non-black things.

Because groups are finite (and if we avoid repeating tests we have already performed), yet another consideration is added to those guiding our intuition: when we are shown many ravens and we see that they are all black, we understand at some point that we have seen a significant percentage of all ravens that exist, and that if there were ravens that are not black, we would have already run into one, because each time we see a black raven, we increase the potential percentage of non-black ravens among those we have not yet checked. By contrast, when we are shown many things that are not black, we know that there are piles, and piles, and piles of non-black things in existence, and that overall we have been shown only a negligible percentage of them; observing such an infinitesimally small percentage cannot convince us that everything that is not black is not a raven.

Chapter 11

Expansions

This chapter contains stories that expand on topics presented in other chapters. Readers who followed the internal links in previous chapters have already read the stories appearing in this chapter.

11.1 Word Power

In "3.1 Think Before You Talk," I noted the importance of the role played by words in human thinking. In this story, I demonstrate the role of words by a problem to which a wordplay provides one of the solutions.

11.1.1 *The problem*

A large rectangle has been tessellated with a finite number of smaller rectangles whose edges are parallel to those of the large rectangle, so that the larger rectangle is covered entirely and there is no overlap between the small rectangles. In each of the small rectangles, the length in centimeters of at least one dimension (height and width) is an integer. Prove that the large rectangle must also have at least one dimension whose length in centimeters is an integer.

11.1.2 *Proof*

Consider a rectangle filled with smaller rectangles, all of which have at least one dimension whose length in centimeters is an integer. Let's remove all the smaller rectangles, but let's remember where each one resided. We will now replace them one by one, starting from the bottom. We will make sure to replace each rectangle only after we have something to "confidently" place it on, so that all the rectangles "supporting" it are already in place, as in the following illustration.

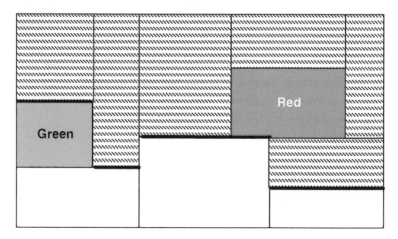

The white rectangles in the above illustration have already been replaced, and the others are still waiting their turn. The green rectangle can already be replaced, but not the red one, because its "support" is not yet entirely ready: before replacing the red rectangle we must replace the hashed rectangle that goes under its right side.

At each stage, we define as the "front" the fragmentary line that describes the highest points marked by the rectangles that have already been replaced. For example, in the illustration above, once the green rectangle has been replaced, the front is marked by the thick black line.

Each point on the front is at a certain distance from the base of the large rectangle. This distance, in centimeters, can be an integer or a fraction. Let's characterize the points at the base of the large rectangle (the "base") as "hard" if the point on the front above them is at a distance of an integer number of centimeters from them, and as "soft" if the front above them is at a distance of a fraction of centimeters above them.

At the beginning of the process, the front is the same as the base of the outer rectangle, and therefore all the points on the base are hard (they are at zero distance from the front, and zero is an integer). Each time we place another rectangle, the characteristic of some of the points on the base may change. The characteristic of a point can change only if we place above it a rectangle with a height that is not an integer number of centimeters, in which case we know that its width is an integer number of centimeters. This means that each time the state of some of the points on the base changes, it happens to a segment of the base that is an integer number of centimeters.

Because at the beginning of the process all the points are hard, and because any change in the nature of the points always occurs along segments whose length is an integer number of centimeters, we can claim that at each stage the size of the set of soft points is an integer: at first it is zero because all the points are hard; later, as it increases or decreases in length, it does so only along segments whose length is an integer number of centimeters.

In the end, after filling the entire outer rectangle, all the base points are at the same distance from the front, which means that they are all either hard or soft. If they are hard, we have our proof because it means that the height of the rectangle is an integer. If they are soft, we also have our proof, because we know that at each stage, including this one, the total length of segments of soft points is an integer, which means that the dimension of the base is an integer because the set of segments makes up the entire base.

It turns out that the problem can be solved simply by... wordplay. In "3.1 Think Before You Talk," I argued that humans are superior to other beasts not so much because they have language, but because they have the ability to invent language. This ability helps humans think by ascribing a name or a symbol to a complex concept, then using that name or symbol to build even more complex concepts and thoughts, and so on. The solution above shows how useful this ability is. Outside the universe of our problem, terms like "front," "base," "hard points," and "soft points" are used in a completely different sense than in the universe of the problem.

And who has ever heard of a hard point, anyway? I could have chosen other words, such as "splanter," "kerming," and "walproe" for the same purpose, but I preferred to use words whose conventional meaning and their meaning within the problem are somehow related.

11.2 Symmetry-Breaking Evolution and a Wild Hypothesis

A few words about evo devo.

In "8.4 Divine Symmetry," I explained that the first multicellular living beings had spherical symmetry, and that the two-sided symmetry we recognize today is merely a remnant of that spherical symmetry, after evolution caused the head to be separated from the tail and the belly from the back. I also explained that the traces of the same evolutionary process are found in the embryonic development of multicellular beings, the beginning of which is always spherical.[166] A more in-depth look at embryonic development and its evolution identifies further and fascinating evidence of the same breaking of symmetry (head, tail, belly, and back).

The hint of the development of differences between head and tail, and its connection to the digestive system, can be found in the process of gastrulation, in which the seeds of the differentiation of cells in the body are sown, and a distinct group of cells forms, from which the digestive system develops. As in the living creature that eventually develops from the embryo, the group of cells of the digestive system is open to the outside world; this opening is the first deviation of the developing embryo from the spherical shape. In this way, a distinct direction is created, called "front," and from the very definition of "front," the definition of "back" is derived.

Evo devo, a nickname for evolutionary developmental biology, is a scientific field that studies the embryonic development processes of various organisms to identify evolutionary connections between them and to understand the evolutionary chain of embryonic development.[167] One of

---

[166] Morula.

[167] A detailed description of the findings of this new science can be found in the book *Endless Forms Most Beautiful* by Sean B. Carroll.

the most exciting discoveries in this study was the development of the asymmetry between the head and the tail, and of the asymmetry between the belly and the back. It turns out that what determines the future head–tail and dorso–ventral directions of the organism is the chemical environment within the mother, but what exactly will develop at each point along the head–tail and the dorso–ventral axes is determined by a component of the genome called "homeotic genes," in particular, Hox genes.[168] These genes control the development of various organs throughout the body, and are arranged in the genome exactly in the order in which the organs controlled by them appear in the complete animal, along the head–tail axis. The phenomenon is present in all animals tested. These genes have been well preserved over the course of evolution, and changes in them in the various animals are relatively few.

Another exciting discovery is the presence of genes that determine whether the tissue that develops from the cells in which they are active will become part of the belly or of the back. These genes are the genetic expression of the breaking of the symmetry between the belly and the back. These genes play opposite roles in vertebrates and in arthropods, so that the gene that plays the "bellying" role in the arthropods serves as the "backing" gene in vertebrates, and *vice versa*. This suggests that arthropods and vertebrates had a common ancestor, one of whose offspring ambulated on his belly, and the other on his back. A more comprehensive and at the same time highly readable description of these discoveries in the field of evo devo can be found in Chapter 12 of Matt Ridley's *Genome*.[169] In this chapter, you will also make the acquaintance of Étienne Geoffroy Saint-Hilaire, who guessed this connection between arthropods and vertebrates as early as 1822, based on observations of the development of embryos, and on the fact that the central nervous system of the arthropods is spread across the abdomen, whereas in vertebrates it is along the back.

So where is the wild hypothesis? The fact that our nervous system is cross-linked, in other words, the left hemisphere of the brain controls the right side of the body, and *vice versa*, together with the belly–back inversion between vertebrates and arthropods may mean that the cross-link is

---

[168] Hox gene.
[169] The entire book is warmly recommended.

the result of the body of the ancestor of the vertebrates having turned in relation to the brain (or, equivalently, the brain having turned within the head).

11.3 Appendix to "8.6 Do We Live in the Matrix?"

Even using nature has not helped solve problems of such complexity that their most efficient computerized solution is not polynomial.

To illustrate this problem, I rely on content from Aaronson's article "NP-complete Problems and Physical Reality."[170] This article presents, among others, a physical experiment in finding Steiner trees using soap bubbles. Steiner trees are defined as follows: Given a collection of points,[171] the Steiner tree of these points is a network of segments on which it is possible to move from any point in the collection to any other point, and where the sum of the length of the segments is minimal (think of a network of roads connecting a group of cities). The network can contain junction points that are not included in the given collection of points, which are called Steiner nodes.

There are polynomial algorithms that, given the topological[172] structure of the network of roads, can find the geometry of the network with this topological structure, where the sum of the lengths of the segments is minimal. It is the need to scan all possible topological structures that precludes a polynomial solution.

Some people expect nature to choose an optimal topological structure, then choose the optimal geometry for this structure, but this is not what happens in practice. Nature seems to arbitrarily choose some topology, then find the optimal geometry for this topology. It does exactly what we can do polynomially.

---

[170] NP-complete Problems and Physical Reality.

[171] The article discusses only points on the plane, but Steiner trees exist also for collections of points in three-dimensional space and, indeed, in spaces with any number of dimensions.

[172] Topology.

11.3.1 *A ray of hope or a conspiracy in spades?*

Steiner trees have several properties that can be used to limit the collection of topological structures they may have. One of the features of Steiner trees is that because of their minimality, the network of segments cannot include closed polygons, because in a closed polygon it is always possible to omit one of the segments without affecting connectivity. In man-made algorithms, we know how to disqualify topological structures that do not comply with these features, and we can restrict the search to appropriate topological structures only. It appears that nature does not do so, and in some cases performs geometrical optimization on a topology that does not fit a minimal network, for example, a topology with a closed polygon.

This finding can be interpreted in several ways, the most interesting among which include the following:

1. The world is not a computer simulation (most of us would look upon this option as a ray of hope).
2. The world is a simulation, but its designers didn't know how to do everything we know (you may try to submit suggestions for optimization to the creator).
3. The world is a simulation, and the above-mentioned failure in the solution of Steiner trees is an attempt to camouflage this fact (conspiracy in spades).

I tend to think that the first option is the correct one. Natural laws operate locally, and not in a goal-oriented manner. In the Steiner tree example, for all practical purposes, the determination which topology is optimized is made under chaotic conditions and, for all practical purposes, is random. This choice cannot be described based on of the macroscopic data of the physical analysis of the soap bubbles.

For an in-depth study of the material of this appendix, I recommend reading the following two articles:

1. Scott Aaronson, "NP-complete Problems and Physical Reality."[173]
2. Arkadi Bolotin, "Computational solution to quantum foundational problems."[174]

---

[173] Scott Aaronson, <u>NP-complete Problems and Physical Reality</u>.

[174] Arkadi Bolotin, <u>Computational solution to quantum foundational problems</u>. Those who are not interested in quantum theory may want to skip the rest of this comment.

One of the most prominent divisions in the <u>interpretations of quantum theory</u> is the one between the interpretations that assume that the wave function collapses when a measurement is performed and those that assume that the wave function does not collapse. Bolotin's theory belongs to the first category. The interpretation accepted by most physicists is the Copenhagen interpretation, which belongs to the first category, but this interpretation does not specify exactly what is a "measurement" leading to collapse. Believers in the first category (and here we are really talking about belief, because today we have no information that can decide between the various interpretations) who are not satisfied with an interpretation that does not define what measurement is, and want to provide a practical definition to the term "measurement," are divided into several groups. One of these groups claims that measurement is created by consciousness, and in the absence of a conscious being who observes the results of the physical process, there is no measurement and no collapse. Only very few scientists accept this claim. Another group proposes an "objective collapse," that is, the existence of defined physical conditions in which the collapse occurs. Among the theories offered in this category, a prominent one is that of Penrose: https://en.wikipedia.org/wiki/Penrose_interpretation. Bolotin's article belongs to this group, and proposes a new and original mechanism of collapse, which defines the conditions of collapse on the basis of complexity. The second category contains several other interpretations, the most prominent and widely accepted of which is that of multiple universes.

Epilogue

The friend I mentioned at the beginning of the book was one of many teens who find religion every year as part of their search for answers to similar questions. He was more fortunate than most, because after 17 years in the ultra-Orthodox community, having a wife and seven children, he set aside religion when he realized that the answers religion purported to provide were false.

Most youths who turn to religion do not find the way back or lack the courage to try. Indeed, it takes great courage, because ultra-Orthodox society is highly intolerant of those who question the tenets of religion, and those who leave the community are often forced to sacrifice all the family ties and friendships they have forged over the years.

I hope this book will help people who face a similar dilemma in the future to understand that it is not necessary to invent God to find our way in the world.

CPSIA information can be obtained
at www.ICGtesting.com
Printed in the USA
BVHW091925290122
627167BV00002B/9

9 789811 230905